THE BACTERIAL CELL SURFACE

THE BACTERIAL CELL SURFACE

**Stephen M. Hammond
Peter A. Lambert and
Andrew N. Rycroft**

CROOM HELM
London & Sydney

KAPITAN SZABO PUBLISHERS
Washington, DC

© 1984 Stephen M. Hammond, Peter A. Lambert and Andrew N. Rycroft

Croom Helm Ltd, Provident House, Burrell Row,
Beckenham, Kent BR3 1AT

Croom Helm Australia Pty Ltd, First Floor, 139 King Street,
Sydney, NSW 2001, Australia

British Library Cataloguing in Publication Data

Hammond, Stephen M.
    The bacterial cell surface.
    1. Bacteria
    I. Title   II. Lambert, Peter Al
    III. Rycroft, Andrew N.
    589.9'0872   QR75

ISBN-13: 978-0-7099-1267-5   e-ISBN-13: 978-94-011-6553-2
DOI: 10.1007/978-94-011-6553-2

Published in the U.S.A. (and Canada)*
by Kapitan Szabo Publishers,
1740 Lanier Place N.W.
Washington, D.C. 20009

Library of Congress Cataloging in Publication Data

Hammond, Stephen M.
    The bacterial cell surface.

    Includes index.
    1. Bacterial cell walls. I. Lambert, Peter A.
II. Rycroft, Andrew N., 1955-   . III. Title.
[DNLM: 1. Bacteria – cytology. 2. Bacteria – physiology.
3. Cell Wall – physiology. QW 52 H227b]
QR77.3.H36 1984   589.9'0875   54-7882

CONTENTS

THE BACTERIAL CELL SURFACE

Stephen M. Hammond is a Lecturer in the Department of Microbiology, University of Leeds

Peter A. Lambert is Lecturer in Microbiology, Department of Pharmacy, University of Aston in Birmingham

Andrew N. Rycroft is Lecturer in Microbiology, University of Glasgow Veterinary School

PREFACE

It is a common statement that because of its simplicity the bacterial cell makes an ideal model for the study of a wide variety of biological systems and phenomena. While no-one would dispute that much of our understanding of biological function derives from the study of the humble bacterium, the concept of a simple life-form would be hotly disputed by any scientist engaged in the determination of the relationship between structure and function within the bacterial cell. Bacteria are particularly amenable to intensive study; their physiology can be probed with powerful biochemical, genetical and immunological techniques. Each piece of information obtained inevitably raises as many questions as answers, and can lead to a highly confused picture being presented to the lay reader. Nowhere is this more evident than in the study of the surface layers of the bacterial cell. Examination of the early electron micrographs suggested that the bacterial cytoplasm was surrounded by some sort of semi-rigid layer, possessing sufficient intrinsic strength to protect the organism from osmotic lysis. The belief that the surface layers were rather passive led to their neglect, while researchers concentrated on the superficially more exciting cytoplasmic components. Over the last twenty years our view of the bacterial envelope has undergone extensive revision, revealing a structure of enormous complexity. In this short monograph we have attempted to explain the known properties of the bacterial envelope and hope to convince the reader that the surface represents a highly integrated series of components that serve as an interface between the metabolic activity of the organism and its environment. In the early chapters we describe the structure and biosynthesis of those layers lying exterior to the cytoplasmic membrane. In the later sections we have attempted to relate the importance of these surface structures to the life of the organism in its natural habitat. At the end of each chapter we suggest some review-type articles which extend and complement the given text.

We are grateful to those people in our respective institutions who have contributed to the production of this book. During the writing of the book we have relied extensively upon the published work of scientists of many disciplines. In attempting to produce a concise, and we hope readable, account we have been unable to acknowledge more than a handful of the research workers involved by name; to these unnamed scientists we dedicate this book.

SMH
PAL
ANR

1 THE PEPTIDOGLYCAN LAYER

The bacterial cytoplasm is enclosed by a rigid, highly structured layer of great mechanical strength, termed the cell wall. The wall is responsible for the maintenance of the shape and integrity of the cell. The cytoplasmic membrane which surrounds the bacterial protoplast has little intrinsic strength and if not overlaid by the wall would be ruptured by the high internal osmotic pressure, if the organism were placed in hypertonic solution. The wall's great strength derives from the unique highly cross-linked aminosugar polymer, called peptidoglycan, of which the wall is built. The importance of the bacterial cell wall to the organism can be demonstrated by treatment with enzymes or certain antibiotics. Many body fluids (e.g. serum, tears) and leucocytes, contain the enzyme lysozyme, which is an important component in the host defences against invading bacteria. Addition of the enzyme to a suspension of a Gram-positive bacterium, e.g. *Bacillus subtilis*, degrades the peptidoglycan, the cells rapidly lyse and the suspension becomes clear and viscous due to the release of the bacterial chromosomal material (DNA) as the cells burst. Lysozyme appears to have unrestricted access to the peptidoglycan of Gram-positive cells. In Gram-negative bacteria the peptidoglycan is masked by a protective layer, the outer membrane (Chapter 3), which prevents lysozyme from penetrating through to the peptidoglycan. Lysis by lysozyme of Gram-negative bacteria can only be achieved by first increasing the permeability of the outer membrane by treatment with a chelating agent, such as ethylenediamine-tetra-acetic acid (EDTA). EDTA exhibits strong binding affinity for metal ions, especially magnesium. In the presence of the organic buffer, tris-hydroxymethylaminomethane, treatment with EDTA withdraws magnesium ions from the outer membrane, leading to the release of specific outer membrane components and greatly increasing the permeability of the membrane. Only under these conditions can lysozyme gain access to the underlying peptidoglycan and bring about cell lysis. Many antibiotics, particularly those possessing a β-lactam ring structure, e.g. penicillins and cephalosporins, bring about their lethal action by inhibiting a number of enzymes involved in peptidoglycan assembly. This leads to a progressive weakening of the cell wall until the structure possesses insufficient strength to resist the internal turgor pressure. Often bulges develop at the site of cell division, which may enlarge until the cells ultimately burst.

Structure of Peptidoglycan

Determination of the composition of bacterial peptidoglycan must be preceded by the preparation of bacterial cell walls free from contamination by other cell constituents. Gram-positive bacteria are susceptible to mechanical breakage, usually achieved by vigorous shaking in the presence of small glass beads in an apparatus such as the Braun Disintegrator. Walls can be readily separated from such a suspension by differential centrifugation and freed from contamination by exhaustive washing.

Approximately 50 per cent of the weight of the Gram-positive cell wall is peptidoglycan, the remainder made up of a variety of polymers depending upon the organism. These accessory polymers, teichoic acids, teichuronic acids and proteins will be discussed in Chapter 2. In order to study the chemical composition of the peptidoglycan, enzymic or chemical treatments are used to cleave the covalent bonds linking the accessory polymers to the peptidoglycan, producing pure peptidoglycan. In Gram-negative organisms, peptidoglycan makes up only 10–20 per cent of the weight of the wall. Although this percentage is much lower than in Gram-positive organisms the peptidoglycan layer is of no less importance to the life of the cell. Purification of Gram-negative peptidoglycan also begins with the mechanical rupture of the cells, usually by extruding the cell suspension through a narrow constriction in a pressure cell, e.g. as in the French Press. The Gram-negative layer is covalently linked by lipoprotein molecules to an outer membrane consisting of phospholipid, protein, lipopolysaccharide and polysaccharide. The peptidoglycan can be readily freed from the majority of the outer membrane components by heating the wall fraction with an anionic detergent such as sodium dodecyl sulphate (SDS), followed by extensive washing to remove the detergent. However, simple detergent-treatment cannot free the lipoprotein which remains covalently attached to the peptidoglycan. Its removal can be effected by treatment with proteolytic enzymes. When stripped of all its associated polymers peptidoglycan is an insoluble macromolecule, which retains the characteristic shape of the original cell from which it was prepared (Figure 1.1).

The chemical composition and structure of the peptidoglycan has been determined for many bacterial species. Although there is considerable variation in detailed composition amongst different organisms, the basic structure is essentially the same. Peptidoglycan is made up of a network of linear polysaccharide chains (glycan strands), up to 200 disaccharide units in length, cross-linked by short peptide chains. The structure of the peptidoglycan found in *Escherichia coli* (Figure 1.2) is typical of most other Gram-negative bacteria and many bacilli (i.e. rod-shaped bacteria). The glycan strands are composed of alternating units of *N*-acetylglucosamine and *N*-acetylmuramic acid joined by a 1,4-β glycosidic linkage

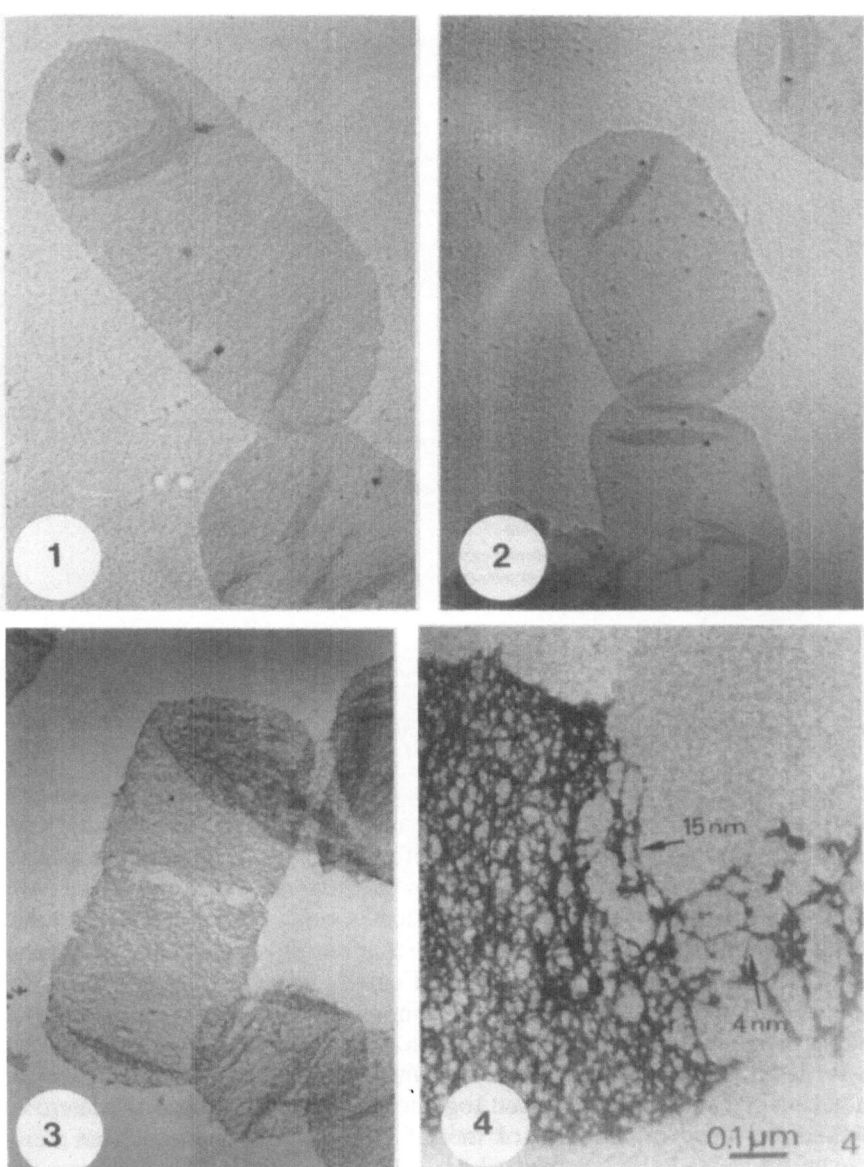

Figure 1.1: Electron Micrograph Showing Isolated Purified Sacculi from *Escherichia coli*. 1 and 2 show the purified sacculi which is in effect one enormous peptidoglycan molecule, 3 and 4 show the sacculi after treatment with specific endopeptidase. It can be seen that the glycan chains appear to run perpendicular to the long axis of the cell. Reproduced by permission, Verwer, R.W., Nanniga, N., Keck, W. and Schwartz, U. (1978) *J. Bacteriol.*, *136*, 723.

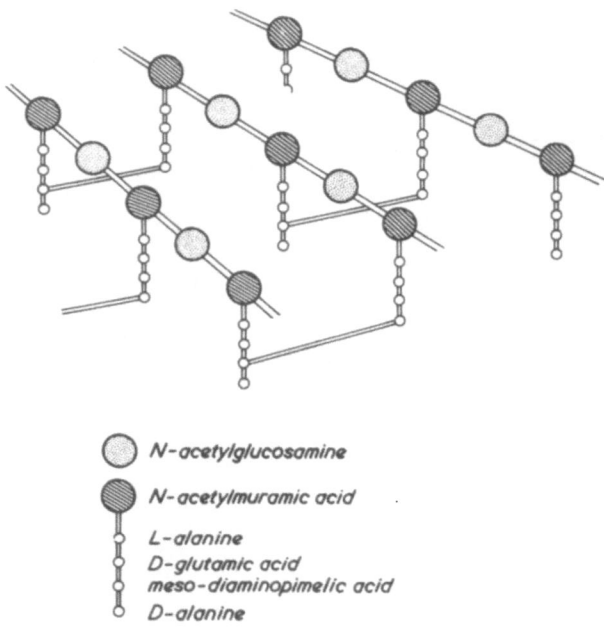

N-acetylglucosamine

N-acetylmuramic acid

L-alanine
D-glutamic acid
meso-diaminopimelic acid
D-alanine

Figure 1.2: Generalised Representation of the Structure of Bacterial Peptidoglycan

(Figure 1.3). *N*-acetylmuramic acid is unique to peptidoglycan and is a derivative of *N*-acetylglucosamine, bearing an ether-linked D-lactyl on carbon 3. To the carboxyl group of each muramic acid residue is attached a chain of four amino acids, joined by a peptide bond. The amino acids have alternating L and D centres, the sequence being L-alanine, D-glutamic acid, meso-diaminopimelic acid and D-alanine. D-amino acids are rare in nature, their occurrence being confined chiefly to peptidoglycan. The third amino acid of the sequence, meso-diaminopimelic acid is also unique to peptidoglycan, and is a symmetrical molecule with two centres of optical activity, one D and one L (hence the name meso). The tetrapeptide chains on adjacent glycan strains are linked together by various means to give a cross-linked polymer. In the case of most Gram-negative bacteria and many bacilli the cross-linking is formed by a direct peptide linkage between the carboxyl group of the fourth amino acid on one glycan chain, D-alanine, and the free amino group of the third amino acid on an adjacent chain, i.e. meso-diaminopimelic acid (Figure 1.4). Similar linkages at other points along the glycan strands build up the cross-linked structure of peptidoglycan. It is thought that the peptidoglycan component of the bacterium is

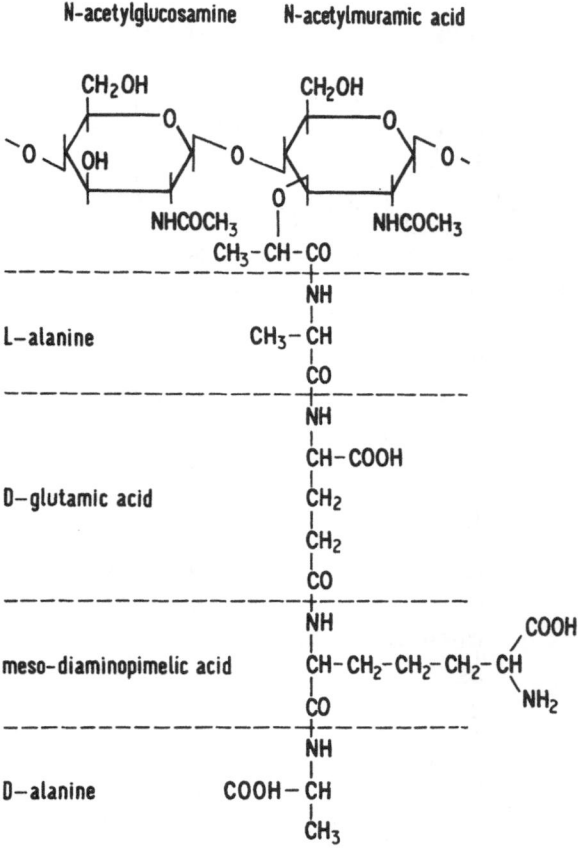

Figure 1.3: Linkage of *N*-acetylmuramic Acid and *N*-acetylglucosamine in the Glycan Strands and Attachment of Amino Acids to Muramic Acid

a single macromolecule covering the entire cell surface. The number of tetrapeptide chains which participate in cross-linking varies considerably from organism to organism. It can be as low as 20 per cent in Gram-negative organisms, such as *E. coli*, with the majority of tetrapeptides remaining unlinked. In contrast, in the Gram-positive organism *Staphylococcus aureus*, the cross-linking may be greater than 90 per cent. In this organism a bridge of five glycine residues links the terminal carboxyl group of D-alanine in one chain to the free amino group of the third amino acid on an adjacent chain, which in this case is L-lysine (Figure 1.5). The

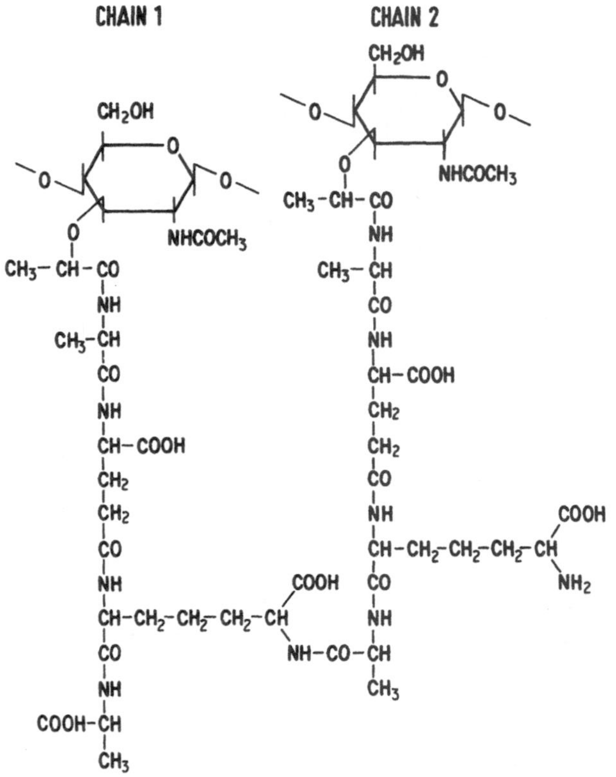

Figure 1.4: Formation of the Peptide Cross-links Between Adjacent Glycan Strands. The linkage is formed by a peptide bond between the ε-NH$_2$ group of diaminopimelic acid on chain 1 and the carboxyl group of the terminal D-alanine on chain 2

peptidoglycan of *S. aureus* has relatively short glycan chains, which necessitates the high degree of cross-linking to produce a structure of equivalent strength to that of other bacteria.

Variations in Composition

Detailed examination of the peptidoglycan from a range of bacteria has revealed many variations in chemical composition, for example the nature of the amino acids joined to muramic acids and forming the cross-links between glycan strands. Attempts have been made to classify peptidoglycans according to the way in which the cross-links are formed. The

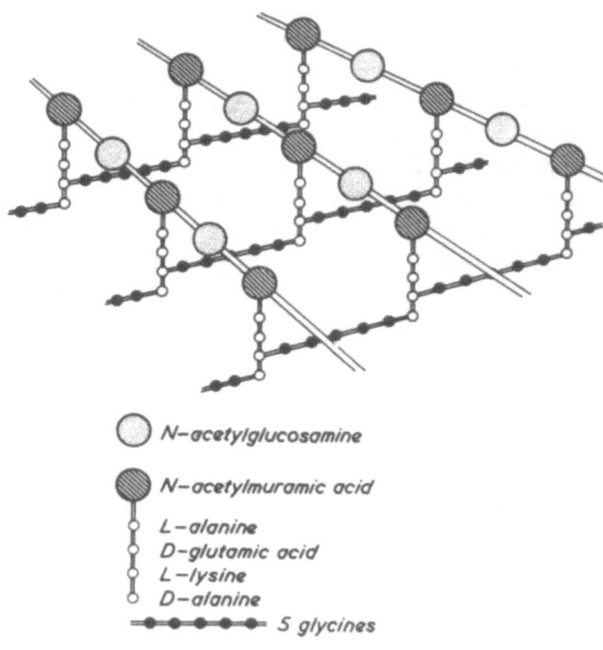

N—acetylglucosamine

N—acetylmuramic acid

L—alanine
D—glutamic acid
L—lysine
D—alanine

5 glycines

Figure 1.5: Structure of Peptidoglycan in *Staphylococcus aureus*

major variation concerns the amino acid at position 3 in the peptide chain on muramic acid. Although meso-diaminopimelic acid is by far the most common occupant of the 3-position, L-lysine is found in many cocci (e.g. *Staphylococcus aureus*, *Micrococcus luteus*, *Sarcina flava*, *Streptococcus faecium*, *Leuconostoc mesenteroides* and *Gaffkya homari*) and some other species contain L-ornithine, L-diaminobutyric acid or L-homoserine. The cross-linkage in most Gram-negative and Gram-positive bacilli is directly formed between the free amino group of the amino acid in position 3 and the terminal carboxyl group of the D-alanine at position 4 on an adjacent glycan strand. In some other organisms an intervening chain of amino acids is present between the glycan chains; for example in *S. aureus* five glycine residues link the two peptide chains on adjacent glycan polymers. In *M. luteus* this bridging polypeptide has the same sequence as the peptide bound to the muramic acid (i.e. L-Ala-D-Glu-L-Lys-D-Ala). In this species a number of muramic acid residues are unsubstituted, suggesting that the bridging polypeptide is derived from tetrapeptides that have been detached from muramic acids.

The D-glutamate residue at position 2 in the muramic acid tetrapeptide is also subject to modification. It should be noted that this moiety is usually linked to the first residue, L-alanine, by its α-amino group (i.e. a normal peptide linkage), but to the third residue through a peptide linkage involving its γ-carboxyl group, leaving the α-carboxyl free. In some species, including *S. aureus, Corynebacterium diphtheriae, Clostridium perfringens* and *Lactobacillus plantarum* the α-carboxyl group of D-glutamate is present as an amide; in *M. luteus* it is modified by addition of a glycine residue and in other species it remains unsubstituted. Diaminopimelic acid also contains a free carboxyl group (at the D centre), which is not involved in any linkage. This group can be unsubstituted (e.g. *E. coli* and *Bacillus megaterium*) or amidated (e.g. *C. diphtheriae, B. subtilis* and *L. plantarum*). In species where lysine occupies position 3 no free carboxyl is present.

Variations also occur in the muramic acid itself. In *S. aureus* approximately half of the muramic acid residues bear an acetyl group on the 6 position. This modification renders the peptidoglycan insensitive to degradation by lysozyme. Similar modifications have been reported in some strains of *Proteus, Neisseria* and *Pseudomonas*. Loss of some of the N-acetyl residues from the N-acetylglucosamine of some *Bacillus cereus* strains also has the effect of rendering the peptidoglycan resistant to lysozyme.

Despite the numerous variations encountered in the peptidoglycan backbone, the peptide side chains and the bridge linking the chains, the overall structure of the peptidoglycan is essentially constant. It forms a cross-linked lattice structure which is responsible for the shape and integrity of the bacterial cell wall (Figure 1.1).

The enzyme lysozyme breaks the linkages between the N-acetylmuramic acid and N-acetylglucosamine units. The structure of lysozyme has been determined by X-ray diffraction and found to contain a cleft, or groove, into which the glycan strand fits perfectly. Once attached to the groove, or active site, the glycosidic bond is broken, significantly weakening the peptidoglycan lattice. The acetylation of muramic acid residues in *S. aureus*, prevents effective binding of lysozyme to the peptidoglycan. However, peptidoglycan of *S. aureus* is sensitive to the enzyme lysostaphin. Lysozyme and lysostaphin are both N-acetylmuramidases, capable of hydrolysis of the N-acetylmuramyl-1,4-β-N-acetylglucosamine bonds in the glycan chains.

A number of enzymes capable of degrading peptidoglycan at other points has been identified. For example, the glycan chains can also be broken by N-acetylglucosaminidase; the linkage between N-acetylmuramic acid and the peptide units is broken by N-acetylmuramyl-L-alanine amidase and the peptide bridges between glycan chains can be hydrolysed by endopeptidase (Figure 1.6). Nearly all bacteria produce enzymes capable of degrading their own peptidoglycan. These autolysins are found

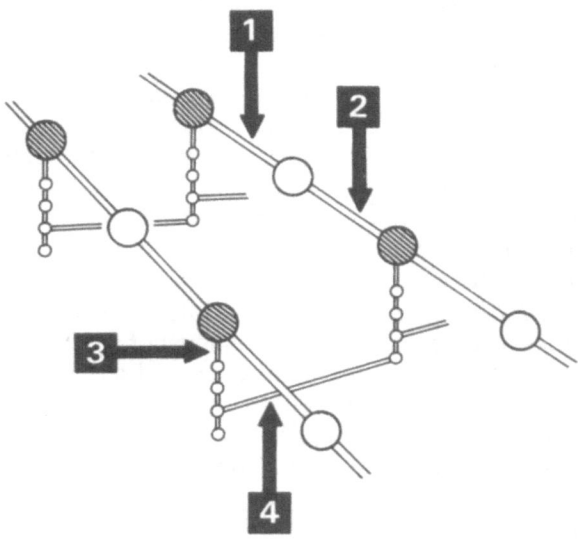

1   *N- acetylmuramidase* (Lysozyme and lysostaphin)

2   *N- acetylglucosaminidase*

3   *N- acetylmuramyl-L- alanine amidase*

4   *DD - endopeptidase*

Figure 1.6: Sites at which Enzymes Attack Peptidoglycan (Symbols for peptidoglycan structure as in Figure 1.2)

in the cell walls of the organism and their function seems to be to modify existing peptidoglycan permitting extension of the wall during cell growth and separation of the cells during division. Clearly their activity must be carefully regulated or these enzymes would rapidly degrade the bulk peptidoglycan ultimately causing cell autolysis. The enzymes are actually bound to, and inhibited by, the polymers in the cell wall. Understanding of the mechanism by which tailoring of existing peptidoglycan and insertion of new material is controlled in the growing bacterial cell in such a way that the daughter cells retain the characteristic shape presents an enormous challenge to microbial biochemists.

Arrangment of Peptidoglycan in the Wall

Surprisingly little firm evidence is available to indicate the way in which peptidoglycan is arranged in the cell wall. The glycan chain length of the peptidoglycan of Gram-positive bacilli is about 100 disaccharide units. This extended polymer represents a length of approximately 0.1 µm. A rod-shaped organism 5 µm long and 0.5 µm wide would require 50 glycan chains to extend the length of the cell, or about 15 to wrap around the circumference of the short axis of the cell. There is some evidence to suggest that in some bacteria, e.g. *B. subtilis* and *E. coli*, the glycan chains are orientated so they wrap around the short axis of the cell, perpendicular to the long axis. Direct observation of *B. subtilis* walls under the electron microscope reveals striations aligned around the short axis circumference of the organism, which persist even after removal of accessory polymers. A similar pattern can be seen in *E. coli* after careful controlled digestion of some of the peptidoglycan cross-links with endopeptidases, thereby loosening the polymer network (Figure 1.1).

Attempts have been made to use X-ray diffraction to determine the exact molecular arrangement of peptidoglycan. Unfortunately the X-ray diffraction patterns produced to date do not give a clear idea of the structure. Suggestions have been made that the glycan chains are arranged in a helix, similar to that of cellulose or chitin, which are linear polymers of poly-$\beta 1 \rightarrow 4$-glucose and poly-$\beta 1 \rightarrow 4$-N-acetylglucosamine respectively, but other data fail to support this view. The difficulties encountered in obtaining sharp X-ray diffraction patterns indicate that the walls do not contain large areas in which the peptidoglycan has ordered repeat structure. This presumably reflects the way in which the polymer is synthesised and inserted into the existing wall during cell growth.

Biosynthesis of Peptidoglycan

Bacteria are faced with the problem of expanding their cell wall in order to grow and divide. In particular, the peptidoglycan matrix must be extended in a controlled way so that the shape and integrity of the cells are preserved. At first sight it is difficult to imagine how the cells achieve this: peptidoglycan forms an enormous cross-linked network which is outside the cytoplasmic membrane and therefore outside the direct metabolic control of the cell. Synthesis of the precursors occurs inside the cell with the utilisation of considerable metabolic energy. How are they assembled, transferred across the cytoplasmic membrane, and finally inserted into the wall?

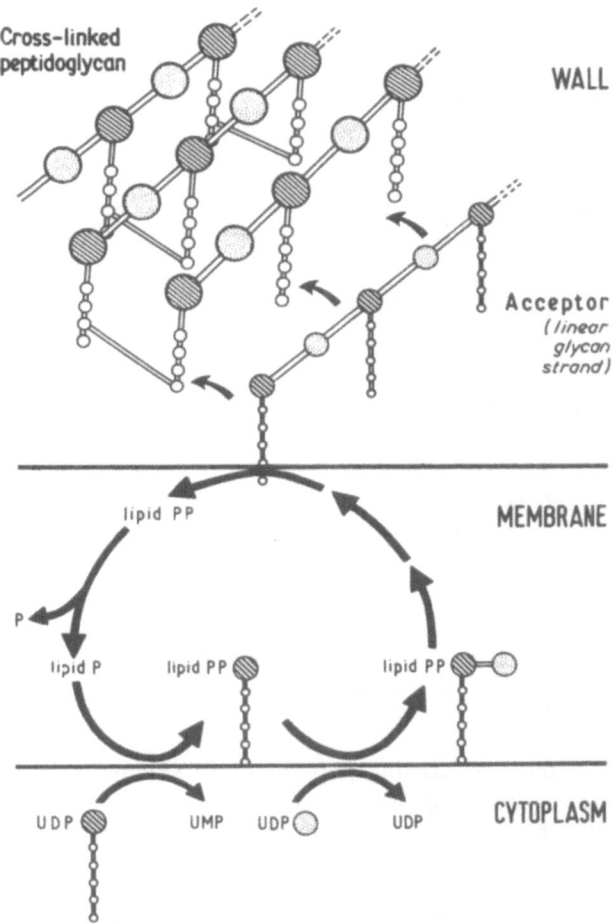

Figure 1.7: Sequence of Peptidoglycan Biosynthesis in *Escherichia coli*. Sugar nucleotides transfer *N*-acetylglucosamine (with its attached pentapeptide) and *N*-acetylmuramic acid to the C_{55}-isoprenoid lipid carrier which serves to translocate the precursor across the cytoplasmic membrane. At the outer face of the membrane the wall precursor is transferred to the growing linear glycan strand, releasing the isoprenoid for another round of synthesis. The acceptor glycan is subsequently cross-linked with existing wall peptidoglycan. (The symbols used for peptidoglycan structure are as in Figure 1.2)

Synthesis of Precursors

The biosynthesis sequence can be divided into three stages: synthesis of the precursors in the cytoplasm; transfer of the precursors to a lipid carrier molecule which transports them across the membrane; insertion into the

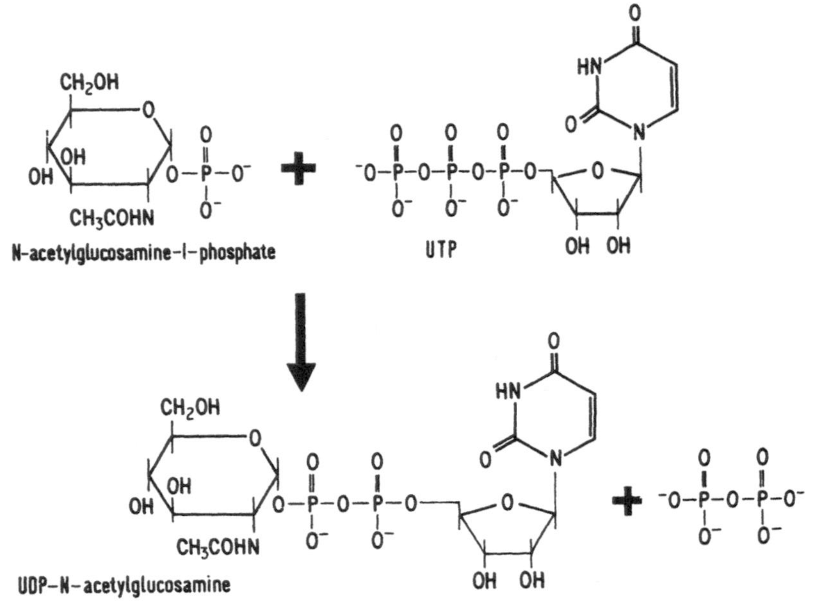

Figure 1.8: Structure of UDP-*N*-acetylglucosamine and its Synthesis from UTP and *N*-acetylglucosamine-1-phosphate

wall and coupling to existing peptidoglycan (Figure 1.7). The precursors synthesised in the cytoplasm are the sugar nucleotides: UDP-*N*-acetylglucosamine and UDP-*N*-acetylmuramylpentapeptide. In nature most polysaccharides are synthesised from sugar nucleotides: the energy for formation of the polymers being provided by hydrolysis of the sugar-nucleotide linkage. UDP-*N*-acetylglucosamine is the precursor of *N*-acetylglucosamine in many polymers and is found in a wide variety of cells. It is synthesised from UTP and *N*-acetylglucosamine-1-phosphate (Figure 1.8) by the enzyme UDP-*N*-acetylglucosamine pyrophosphorylase. By contrast, *N*-acetylmuramic acid is unique to peptidoglycan and its precursor, UDP-*N*-acetylmuramylpentapeptide is used exclusively by bacteria. It is synthesised from UDP-*N*-acetylglucosamine by addition of a 3-carbon fragment from phosphoenolpyruvate to the 3-position of *N*-acetylglucosamine (Figure 1.9). The pyruvyl group is reduced by an NADPH-reductase enzyme giving UDP-*N*-acetylmuramic acid, the 3-*O*-D-lactyl ether of UDP-*N*-acetylglucosamine.

To function as the precursor of peptidoglycan UDP-*N*-acetylmuramic acid is converted into the pentapeptide. A chain of five amino acids is

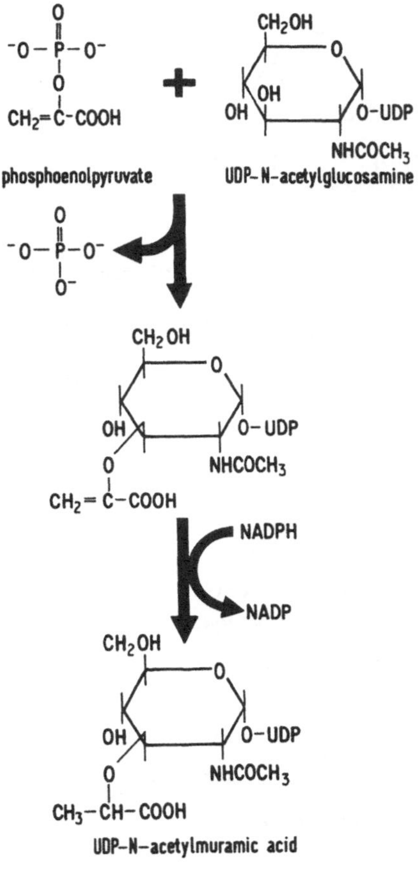

Figure 1.9: Structure of UDP-*N*-acetylmuramic Acid and its Synthesis from UDP-*N*-acetylglucosamine and Phosphoenolpyruvate

linked by an amide bond to the free carboxyl of the lactyl ether substituent, giving the sequence: *N*-acetylmuramic acid-L-alanine-D-glutamic acid-meso-diaminopimelic acid-D-alanine-D-alanine. Note that at this stage the precursor has an extra D-alanine residue which is removed during incorporation into peptidoglycan leaving the familiar tetrapeptide. The first three amino acids are added sequentially to UDP-*N*-acetylmuramic acid by specific enzymes which use ATP as an energy source. The remaining two D-alanine residues are added together as a pre-formed dipeptide, D-alanyl-D-alanine (Figure 1.10). The dipeptide is synthesised from L-alanine by alanine racemase and D-alanyl-D-alanine ligase. The first enzyme converts two molecules of L-alanine into the D forms and the

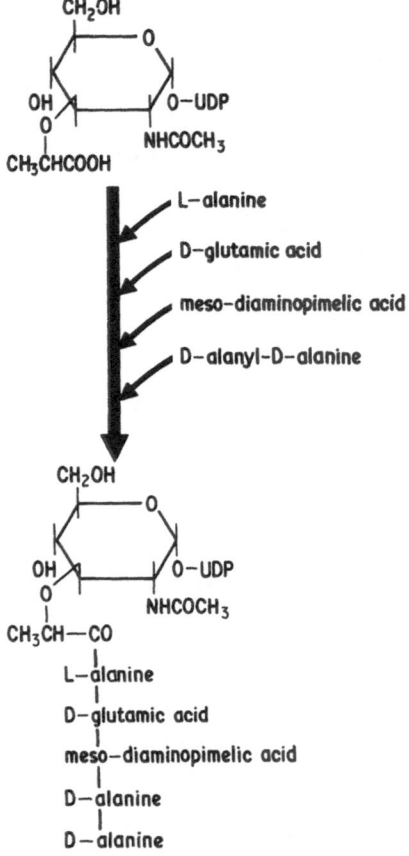

Figure 1.10: Sequence of Addition of Amino Acids to UDP-*N*-acetylmuramic Acid

second joins them together. The terminal D-alanyl-D-alanine portion of *N*-acetylmuramylpentapeptide plays a crucial part in the final cross-linking stage of peptidoglycan synthesis.

The Lipid Carrier

The second stage occurs in the cytoplasmic membrane and is concerned with transporting the precursors from the cytoplasm to the wall. The key molecule responsible for the translocation is the lipid carrier, undecaprenyl phosphate (Figure 1.11). This highly lipophilic molecule is a C_{55}-iso-prenoid alcohol phosphate. It is involved in the synthesis of several other bacterial wall components, for example the O side chains of lipopoly-

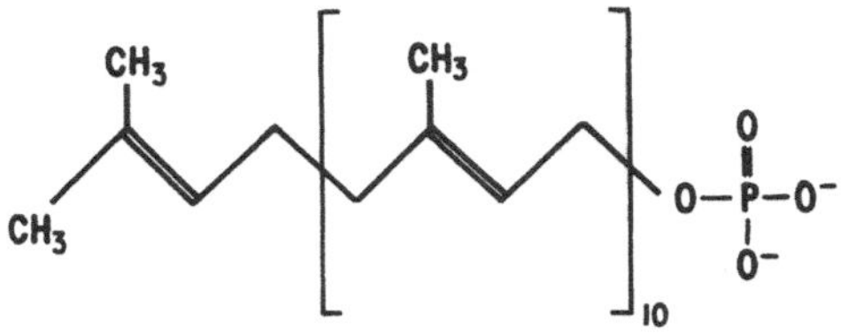

Figure 1.11: Structure of the Lipid Carrier Isoprenoid, Undecaprenyl Phosphate

saccharide, capsules, exopolysaccharides, teichuronic acids, and the linkage unit that joins teichoic acids to peptidoglycan. As a common participant in several biosynthetic sequences it is in a position to regulate the synthesis of wall components. Although undecaprenyl phosphate acts as the acceptor for the precursors of peptidoglycan, the free isoprenoid alcohol also exists in many bacterial membranes. In order to function as the carrier lipid it must first be phosphorylated by another membrane-bound enzyme, undecaprenyl phosphate phosphokinase.

The sequence of events by which the precursors are transferred to the lipid carrier is as follows. Firstly, the phosphoryl-*N*-acetylmuramylpenta-peptide is transferred from UDP-*N*-acetylmuramylpentapeptide to the lipid carrier with release of the nucleotide as UMP; *N*-acetylmuramyl-pentapeptide is now linked by a pyrophosphate bond to the lipid carrier. Secondly, *N*-acetylglucosamine is transferred from UDP-*N*-acetylglucos-amine to the lipid phosphoryl-*N*-acetylmuramylpentapeptide with release of UDP. *N*-acetylglucosamine is glycosidically linked to the muramic acid on the lipid carrier to form a $\beta 1 \rightarrow 4$-linked disaccharide pentapeptide (Figure 1.12). The enzymes responsible for the transfer of the two pepti-doglycan precursors from their nucleotides to the lipid carrier, a translocase and a transferase respectively, presumably act on the inner face of the cytoplasmic membrane. They each utilise both hydrophobic and hydro-philic substrates, i.e. the undecaprenyl phosphate carrier lipid and the sugar nucleotides. When extracted from the cytoplasmic membrane using organic solvents or detergents, the activity of both enzymes is stimulated by the addition of lipids.

In some bacteria the disaccharide pentapeptide, linked to the carrier lipid, can be directly transferred across the cytoplasmic membrane and inserted into the existing peptidoglycan at a growing point in the cell wall. However, in other species certain modifications and additions are pre-requisites for transfer through the membrane. In *S. aureus* the completed peptidoglycan is cross-linked by a bridge of five glycine units and the carboxyl group of the second amino acid in the peptide substituent of

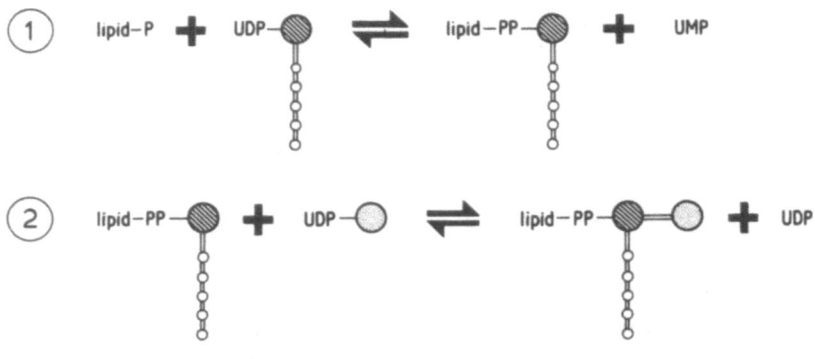

Figure 1.12: Transfer of Muramyl Peptide and *N*-acetylglucosamine from Their Nucleotide Precursors to the Lipid Carrier

muramic acid is amidated. These modifications are all made before insertion into the wall and presumably before the precursors are transported across the cytoplasmic membrane. An interesting feature of the addition of the five glycine units is that transfer ribonucleic acid (tRNA) is involved. The first glycine is transferred from glycyl-tRNA to the amino group of the third amino acid of the peptide side chain (in *S. aureus* this is lysine). A chain of five glycine residues is built up by four subsequent glycine additions to the amino terminus of the growing peptide by glycyl-tRNA. Such a mechanism is quite distinct from the addition of amino acids during protein synthesis, where the growing peptide is linked to tRNA as a peptidyl-tRNA intermediate and each new amino acid is added to the carboxyl terminus of the growing peptide chain. Modifications of the petidoglycan precursors are presumably made at the inner face of the cytoplasmic membrane, a region accessible to both the enzymes and the precursors.

Once the lipid intermediate has transported the disaccharide pentapeptide across the cytoplasmic membrane, from the inner to the outer face, the peptidoglycan precursor is transferred to an acceptor molecule. Although the precise details are still unkown there is evidence to suggest that the acceptor is a linear glycan strand, linked by a pyrophosphate group to a molecule of the lipid carrier. Addition of the disaccharide pentapeptide appears to result from transglycosylation at the reducing end of the growing glycan chain, i.e. the muramic acid residues become joined to the lipid carrier (Figure 1.13). In this way linear glycan strands of about 10 disaccharides are produced. The molecule of lipid pyrophosphate released

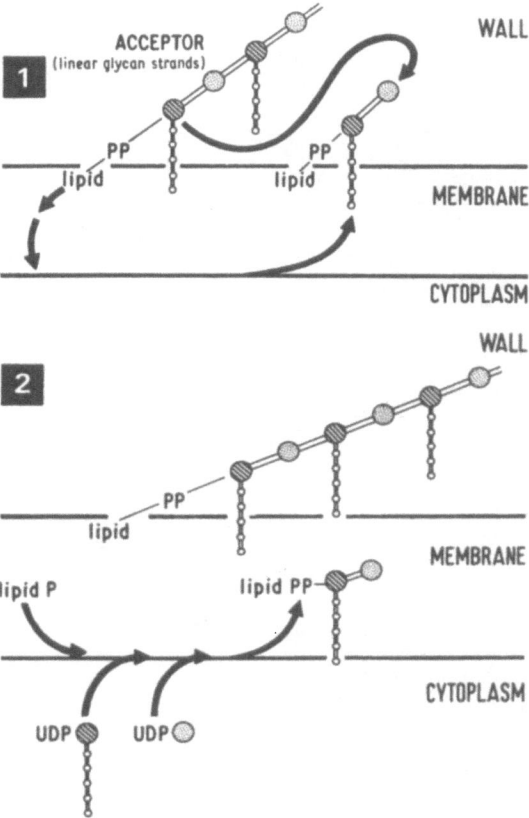

Figure 1.13: Transfer of Lipid-disaccharide Pentapeptide to the Acceptor Glycan in the Wall

at this stage can re-enter the cycle once a phosphate group has been removed by a pyrophosphatase. It is then able to accept another disaccharide pentapeptide unit on the inner face of the cytoplasmic membrane (Figure 1.7).

Cross-linking

This third and final stage of peptidoglycan biosynthesis concerns the way in which the linear glycans, assembled on the lipid carrier at the outer face of the cytoplasmic membrane, become incorporated into the wall. Bearing in mind the final cross-linked structure of peptidoglycan (Figures 1.2 and 1.5), it is clear that new peptide bonds must be made between newly synthesised (nascent) glycan strands and existing peptidoglycan. The reaction is called transpeptidation and is carried out by enzymes known as

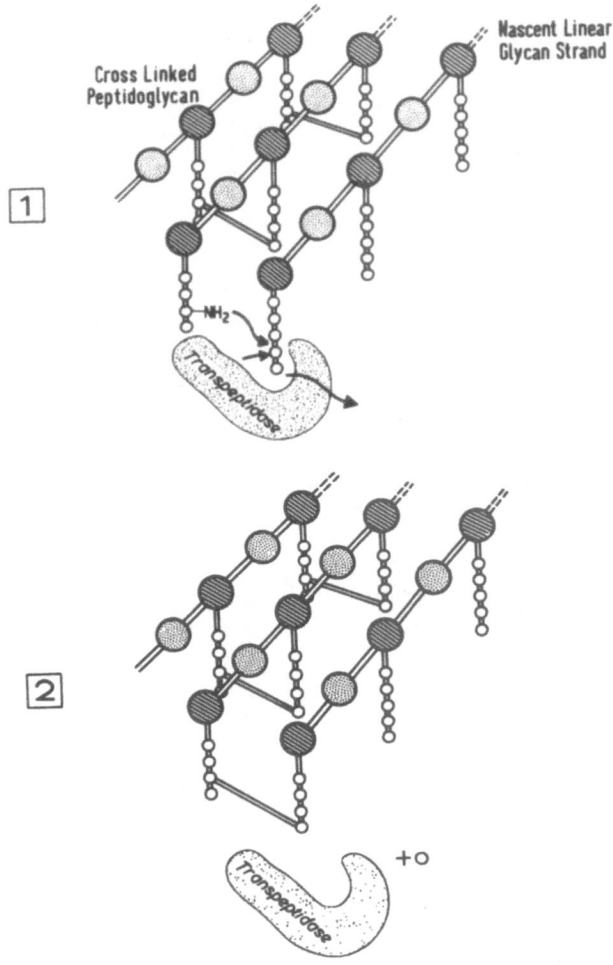

Figure 1.14: Cross-linking of the Nascent Glycan Strand to the Cell Wall Peptidoglycan by Transpeptidase Enzyme in *Escherichia coli*. (Symbols as in Figure 1.2)

transpeptidases. The sequence of events leading to formation of cross-links in *E. coli* has several distinct stages. The transpeptidase first recognises and binds to the terminal D-alanyl-D-alanine unit of a pentapeptide chain linked to muramic acid on a nascent glycan strand. The peptide bond between the D-alanine residues is broken and the terminal alanine released. A new peptide bond forms between the carboxyl group of the remaining D-alanine and the free ε-amino group on a diaminopimelic acid residue on a nearby region of peptidoglycan. The nascent glycan thereby becomes linked to the existing peptidoglycan of the wall; similar trans-peptidations serve to cross-link the glycan strands at other points and firmly knit the network together (Figure 1.14). Explanation has been sought as to the purpose of the fifth amino acid (D-alanine) present on the side chain of the nascent glycan, since it is excised from the final structure. Since the cross-linking reaction occurs within the cell wall, it is necessarily remote from conventional sources of metabolic energy, such as ATP. It appears that the energy released from breaking the terminal D-alanyl-D-alanine linkage is essential for the formation of the peptide cross-link. The peptidoglycan of *E. coli* and many bacilli is only 20–30 per cent cross-linked, yet the remaining peptides which are not cross-linked are all tetra-peptides, lacking the fifth D-alanine. An enzyme, D,D-carboxypeptidase, is responsible for removing the terminal D-alanine from the pentapeptides on nascent linear glycan chains. The activity of this enzyme is very similar to that of transpeptidase, in that it binds to the D-alanyl-D-alanine portion, cleaves the peptide linkage and releases the terminal D-alanine. However, instead of carrying out the transpeptidation, the carboxypeptidase leaves the peptide as an uncross-linked tetrapeptide, terminating in the carboxyl group of D-alanine.

If the carboxypeptidase acts upon the peptide before the transpeptidase then it would effectively regulate the degree of cross-linking by denying the transpeptidase its substrate. Very little is known about the regulation of transpeptidase and carboxypeptidase enzymes, but organisms such as *S. aureus* which have highly cross-linked peptidoglycan have very low carb-oxypeptidase activity. It therefore seems that the activities of the enzymes combine to control the cross-linking. Transpeptidases and carboxypep-tidases are different molecular species, although in its action, carboxy-peptidase is an uncoupled form of transpeptidase, transferring the pen-ultimate D-alanine to water rather than an amino group of an adjacent peptide.

Other enzymes are also involved in the incorporation of glycan strands into the wall. An endopeptidase is capable of breaking cross-links in pepti-doglycan, i.e. it reverses the action of transpeptidase. Its function is prob-ably to create sites in existing peptidoglycan at which new glycan strands can be incorporated by subsequent transpeptidase action. This activity would provide local sites for peptidoglycan expansion and wall extension. Such sites might be important at the point of cell division where consider-able modification to the cell wall, both degradative and synthetic, must occur to allow the daughter cells to separate.

Present information indicates that the nascent chains assembled on carrier lipids on the outer face of the cytoplasmic membrane are only about 10 disaccharides long (although in cell-free preparations they can be considerably longer). This suggests that transglycolase enzymes, capable of extending the glycan chains by inserting nascent glycan strands, must play a central role in producing the fully developed glycan strand of between 100 and 150 disaccharide units.

Clearly the final stage of peptidoglycan synthesis is very complex. It involves a group of enzymes: transpeptidases, carboxypeptidases, probably transglycolases, endopeptidases, amidases and muramidases, which effectively combine to refashion the peptidoglycan network during cell growth. Wall synthesis involves both degradation of existing peptidoglycan, incorporation of new material and the linking of new material to the existing wall polymers. All of this occurs outside the cytoplasmic membrane and therefore outside the direct metabolic control of the cell and yet it occurs efficiently and in such a way that the characteristic shape of the organism is retained during growth and division.

Antibiotics which Affect Peptidoglycan Synthesis

A wide range of antibiotics are known to inhibit peptidoglycan synthesis. Their sites of action have, in many cases, been identified and it is clear that all stages of the peptidoglycan biosynthesis pathway are vulnerable to inhibitors. A selection of such compounds is listed in Table 1.1, arranged in the order in which they affect the sequence (starting from the synthesis of precursors in the cytoplasm), without reference to any clinical usefulness. Probably the most important feature of these agents is that they owe their selective toxicity, i.e. their ability to kill or inhibit bacteria without damaging mammalian cells, to the fact that the peptidoglycan has a unique structure and is vitally important to bacteria. Nearly all bacteria have an absolute requirement for peptidoglycan in their walls, whereas mammalian cells do not possess cell walls, they are bounded solely by their cytoplasmic

Table 1.1: Some Antibiotics which Interfere with the Synthesis of Peptidoglycan

Antibiotic	Site of Action
phosphonomycin	inhibits synthesis of UDP-*N*-acetylmuramic acid from UDP-*N*-acetylglucosamine and phosphoenolpyruvate
D-cycloserine	inhibits conversion of L-alanine into D-alanine by alanine racemase and coupling of two D-alanines by D-alanine: D-alanine ligase
bacitracin	prevents de-phosphorylation of undecaprenyl pyrophosphate thereby inhibiting the lipid carrier cycle
vancomycin	binds to acyl-D-alanyl-D-alanine region of the lipid-linked precursors preventing incorporation into the wall
β-lactam antibiotics e.g. penicillins and cephalosporins	prevent cross linking by inhibition of transpeptidase/carboxypeptidase

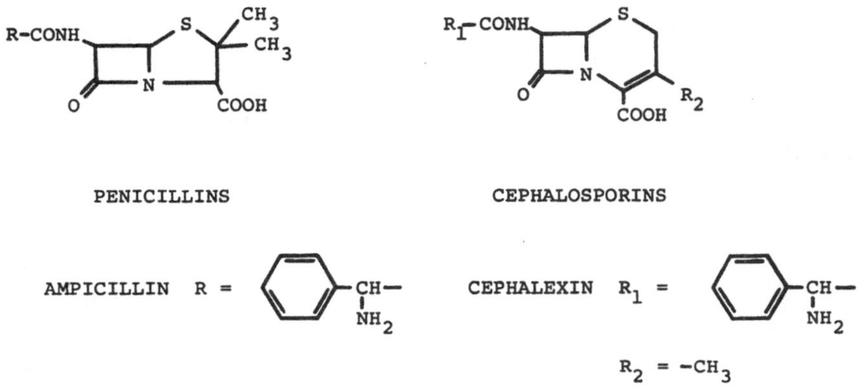

Figure 1.15: The Structure of Penicillins and Cephalosporins, the Classical β-lactam Antibiotics

membrane and do not contain any molecules resembling peptidoglycan. It is the unusual chemical composition of peptidoglycan which makes it such a good selective target for antibiotic action. A number of its components are unique to peptidoglycan: muramic acid, the D-forms of the amino acids, and meso-diaminopimelic acid. Enzyme inhibitors which are analogues of these components or of intermediates in their synthesis are therefore good candidates for selective inhibition of peptidoglycan synthesis and selective antimicrobial action.

The agents listed in Table 1.1, along with many others have played important parts in unravelling the biosynthetic pathway. Without doubt, the most important group of antibiotics, both clinically and as biochemical probes, are the β-lactams; principally the penicillins and cephalosporins (Figure 1.15): These agents, more than any others, have provided the key to our present understanding of the final stage of the process, the cross-linking reaction. β-lactams derive their name from the four-membered cyclic amide ring. In the penicillins, the β-lactam structure is fused with a five-membered ring containing sulphur (a thiazolidine ring), and in the cephalosporins it is fused with a six-membered ring, also containing sulphur (a thiazoline ring). Many semi-synthetic derivatives have been produced by varying the substituents on the basic penicillin and cephalosporin nuclei. The result has been the introduction of a large number of β-lactam antibiotics for the treatment of bacterial diseases. These 'classical' semi-synthetic β-lactams have recently been joined by a number of novel 'non-classical' β-lactams (Figure 1.16). Mecillinam is a penicillin with an

**Mecillinam (an
amidino penicillin).**

**Cefoxitin (a
cephamycin).**

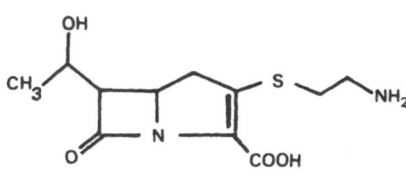

$R_1 = $ thiophene $-CH_2-$

$R_2 = -CH_2OCONH_2$

Clavulanic acid.

Thienamycin.

Nocardicins.

Monobactams.

$X = H, CH_3$

Figure 1.16: The Structure of Some Novel β-lactams

unusual amidino side group; cefoxitin is a cephamycin, a cephalosporin
with an additional methoxy substituent; clavulanic acid contains an unsubstituted β-lactam ring fused with a five-membered ring containing
oxygen in place of sulphur; thienamycin is a penicillin-like molecule in
which the sulphur atom is on a side chain. Finally there are two groups of
compounds containing β-lactam rings which are not fused with a second
ring system, the nocardicins and the monobactams. It is interesting to note

that these β-lactam nuclei are produced naturally by a wide range of micro-organisms: the penicillins and cephalosporins by fungi; clavulanic acid, thienamycin and nocardicin by Actinomycetes, and the monobactams by bacteria. Presumably in the natural environment these organisms derive a protective advantage against competing bacteria by inhibiting their growth through production of β-lactams.

How do β-lactams inhibit the synthesis of peptidoglycan? From the earliest studies with benzyl penicillin (pencillin G) in the 1940s it was realised that synthesis of the bacterial cell wall was affected. Fleming, who was an expert on lysozyme, noted that *S. aureus* colonies growing in the vicinity of a contaminating fungal colony on a culture plate were destroyed. The cells appeared to be lysed by metabolites, i.e. penicillins, released from the fungal colony in a manner reminiscent of the action of the enzyme lysozyme. Later studies showed that lysis only occurred if the cells were actively growing and that sugar nucleotides, subsequently shown to be precursors of peptidoglycan, accumulate in the cytoplasm of growing cells exposed to penicillin. Pronounced effects upon the shape of bacteria, such as bulges, filaments and swollen round forms were also noted depending upon the organism and the conditions. These observations all pointed to an effect upon the peptidoglycan as the component responsible for the shape and integrity of the wall. Since these early days understanding of the mode of action of β-lactam antibiotics has proceeded together with an under-standing of the biosynthesis of peptidoglycan.

When bacteria are exposed to β-lactams a small amount of the anti-biotic becomes covalently linked to the cells. The amount bound varies with the nature of the organism and the antibiotic but it is always very low, of the order of 200–4000 molecules per cell. These covalently bound molecules are responsible for damaging and, ultimately, killing the cell. They are bound to transpeptidases and carboxypeptidases and effectively inhibit the enzymes. Since no other metabolic or biosynthetic activities are affected the cells continue to grow but are unable to produce cross-linked peptidoglycan. Abnormal cell shapes develop and the cells eventually stop growing. The precise fate of the cells depends upon the nature and the concentration of the β-lactam involved and upon the organism itself. Lysis is a common phenomenon but filaments and cells with other unusual shapes can persist without lysis.

The mechanism by which penicillin becomes covalently linked to the transpeptidases and carboxypeptidases is illustrated in Figure 1.17. The β-lactam bond of the antibiotic is broken by the enzyme forming an inactive penicilloyl-enzyme intermediate. The intermediates are sufficiently stable to ensure that all of the transpeptidase and carboxypeptidase molecules in the cell are inactivated and cross-linking of the peptidoglycan cannot occur. Enzyme activity is only slowly restored as the penicilloyl-enzyme complexes break down yielding penicillin fragments. It may not be

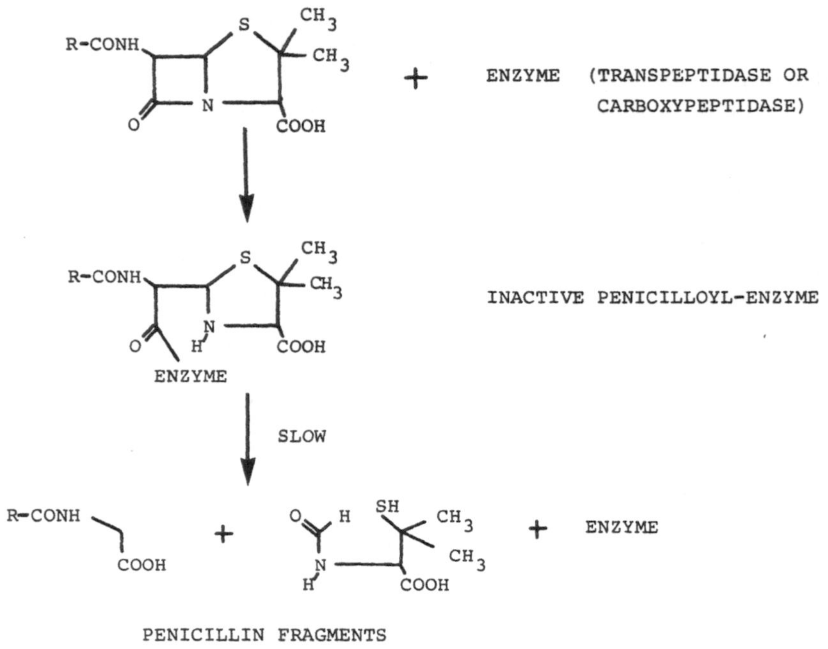

Figure 1.17: Mechanism of Inhibition of Transpeptidase or Carboxypeptidase by Penicillins

immediately obvious why the enzymes bind the β-lactams. A number of explanations have been made, each based on structural similarities between the antibiotics and various portions of peptidoglycan. The most acceptable suggestion at present is that the β-lactams are structural analogues of the D-alanyl-D-alanine portion of nascent peptidoglycan, i.e. the terminal portion of the pentapeptide linked to *N*-acetylmuramic acid in uncross-linked peptidoglycan. The initial action of both transpeptidase and carboxypeptidase is to bind to D-alanyl-D-alanine and break the peptide bond. The reactive and susceptible β-lactam bond of the antibiotic molecule lies in a similar position as the peptide bond of D-alanyl-D-alanine. It is reasonable to suppose that the enzymes mistake the antibiotic molecules for their genuine substrate; cleave the β-lactam bond, and, in so doing, are rendered inactive due to the relative stability of the antibiotic-enzyme intermediate produced. In fact, the bond angles and lengths of the β-lactams and D-alanyl-D-alanine do not correspond as closely as we might expect; it is more likely that the structural similarity is between the antibiotic and a conformation adopted by the substrate when it interacts with the enzymes.

Recognition that β-lactams become covalently linked to their target enzymes has led to considerable interest in studying the relatively stable enzyme-antibiotic complexes, or penicillin binding proteins (PBPs) pro-duced. Using penicillin G containing a radioactive isotope [^{14}C] it is possible to detect the very small quantities of PBPs after treatment of whole cells or membranes with the labelled antibiotic. The usual procedure is to separate the PBPs by electrophoresis on polyacrylamide gels in the presence of the anionic detergent, sodium dodecyl sulphate. Individual PBPs separated in this way according to their size are detected by fluoro-graphy; a scintillant is incorporated into the gel to convert the β-emission from the [^{14}C]penicillin-labelled PBPs into light, which can be detected on an overlying photographic film. The enhancement of the labelled bands achieved by fluorography, carried out if necessary over long exposure periods, permits the detection of extremely small amounts of labelled proteins. When the technique is applied to an organism such as *E. coli*, either with whole cells or isolated membranes, a surprising result is obtained. There are not just two PBP bands, representing the carboxy-peptidase and transpeptidase enzymes present in the organism, but a set of at least seven different PBPs (Table 1.2) with molecular weights ranging from 40,000 to 90,000. The most abundant PBPs are those of lowest molecular weight; PBPs 5 and 6 account for 90 per cent of the total penicillin bound by the cells (i.e. ~ 2,000 molecules/cell). They have carboxypeptidase activity and, whilst they possibly control the degree of cross-linking of the peptidoglycan, they are not considered to be essential for the viability of the cells. The same applies to PBP4, which has been shown to exhibit carboxypeptidase and endopeptidase activity in model systems. The other PBPs are almost certainly essential to the cells; if they are damaged or inhibited then the cells suffer a variety of consequences resulting from impaired ability to assemble peptidoglycan in the correct, ordered fashion. PBP1b has transpeptidase and transglycosylase activity, the activities of the remaining PBPs 1a, 2 and 3 have not yet been estab-lished. It seems that 1a, 1b, 2 and 3 have distinct functions within the cells, each one being responsible for different enzymic activities required to produce peptidoglycan in a form which maintains the rod shape of the organism. PBPs 1a and 1b seem to play vital parts in maintaining the integrity of the cells and possibly in controlling extension of the pepti-doglycan in the side walls during cell growth. Selective inhibition of 1b leads to lysis of the cells. PBP2 is also involved in maintaining the rod shape of *E. coli*; selective inhibition of its enzymic activity leads to the formation of oval-shaped cells which are osmotically stable, i.e. the cells do not lyse. PBP3 is involved in septum formation; selective inhibition of its activity results in long filaments of cells which lack dividing septa. The filaments are not osmotically fragile and do not normally undergo lysis.

Pencillin G binds to all of the PBPs described; that is how they were discovered and how they are defined. Which of PBPs is the major killing target when cells are exposed to penicillin G depends upon the concentration employed but is most likely to be 1b, since the cells are observed to undergo rapid lysis. However, other β-lactam antibiotics do not necessarily have the same affinities for the PBPs as pencillin G. This has been established using either different labelled antibiotics or by carrying out experiments in which unlabelled antibiotics compete with labelled pencillin G for the PBPs. Some surprising facts have emerged from these studies. For example, cephalexin has a selective affinity for PBP3 at low concentrations, and under these conditions its major effect upon the cells is to produce filaments. Mecillinam has a marked selective affinity for PBP2 and its effect upon the cells is to induce the formation of osmotically stable oval forms. Unlike mecillinam, which has little affinity for the other PBPs, most β-lactam antibiotics bind to all of the PBPs but exhibit some preference for certain PBPs at low concentrations. Not surprisingly then, their morphological effects depend upon the concentration employed, and can vary between the extremes of lysis and filamentation with a combination of effects resulting in abnormal cell shapes. Apart from alterations in shape resulting from interference with the biosynthesis of peptidoglycan, the triggering of various autolytic enzymes in the wall by β-lactam antibiotics also plays an important part in the eventual destruction of the cell.

Most of the detailed information on PBPs has been gained by studying *E. coli*. Similar patterns are seen with other Gram-negative bacteria and it is reasonable to suppose that their PBPs have similar functions, although the individual molecular weights may vary slightly. Gram-positive rod-shaped organisms also have multiple PBPs but the patterns obtained for Gram-positive cocci are generally simpler with fewer individual PBPs.

Bacteria Lacking a Peptidoglycan Layer

The peptidoglycan polymer is not present in all members of the bacterial kingdom. It is possible to identify two important groups of bacteria which do not possess a true cell wall.

Table 1.2: The Penicillin Binding Proteins (PBPs) of *E. coli*

PBP	Relative molecular Mass (M_r)	Molecules per cell	Function	Result of inhibition
1a	91,000		?	?
1b	86,000–81,000	230	integrity	lysis
2	66,000	20	shape	oval forms
3	60,000	50	division	filaments
4	49,000	110	endopeptidase/ carboxypeptidase	none
5	42,000	1800	carboxypeptidase	none
6	40,000	570	carboxypeptidase	none

Mycoplasmas. Members of this group are small prokaryotic organisms, some of which are pathogenic to man and animals. It is possible to divide mycoplasmas into two genera, those requiring sterols for growth (*Mycoplasma*) and those which do not (*Acholeplasma*), although the latter will incorporate sterols if an exogenous supply is available. Although mycoplasmas do not possess a demonstrable peptidoglycan layer they are resistant to osmotic lysis, indeed membranes isolated from mycoplasmas are far more robust than other bacterial membranes. This may be a direct consequence of the presence of sterols in the cell membranes.

Archebacteria. The name of this group of microorganisms derives from its first described members, the anaerobic methanogenic bacteria possessing a unique metabolism based on the reduction of carbon dioxide to methane. The apparent antiquity of the organism's biochemistry, plus the fact that they would seem well suited to life in the primordial environment, led to the coining of the term archebacteria. An important difference between the archebacteria and the remaining members of the bacterial kingdom is their unique 16S ribosomal RNA. When it was discovered that certain halophilic bacteria (halophile = 'salt-loving') also possessed similar 16S RNA they were also included in the archebacteria.

Neither the methanobacteria nor the halophiles possess detectable levels of peptidoglycan. In its place the walls of methanobacteria contain a polymeric layer made up of a sugar-substituted polypeptide associated with a polysaccharide containing neutral and amino sugars. The main wall polymer of *Halococcus* is a sulphated heteroglycan containing glucose, mannose and galactose plus galacturonic, glucuronic and gulosaminuronic acids. The lipids of archebacteria are markedly different from other bacteria. The properties of these lipids are discussed in Chapter 6 in the context of their role in the natural habitat of the organism (p. 212).

Further Reading

Burge, R.E., Adams, R., Balyuzi, H.H.M. and Reaveley, D.A. 'Structure of the Peptidoglycan of Bacterial Cell Walls II', *Journal of Molecular Biology* (1977), *112*, 955–74

Burge, R.E., Fowler, A.G. and Reaveley, D.A. 'Structure of the Peptidoglycan of Bacterial Cell Walls I' *Journal of Molecular Biology* (1977), *112*, 927–53

Formanek, H., Formanek, S. and Wawra, H. 'A Three-dimensional Atomic Model of the Murein Layer of Bacteria', *European Journal of Biochemistry* (1974), *46*, 279–94

Gale, E.F., Cundliffe, E., Reynolds, P.E., Richmond, M.H. and Waring, M.J. *The Molecular Basis of Antibiotic Action* 2nd edn (John Wiley and Sons, London, 1980), pp. 79–174

Ghuysen, J.M. *The Bacteria DD-carboxypeptidase-transpeptidase Enzyme System: a New Insight into the Mode of Action of Penicillin* (University of Tokyo Press, 1977)

Ghuysen, J.M. 'The Concept of the Penicillin Target from 1965 Until Today', *Journal of General Microbiology* (1977), *101*, 13–33

König, H. and Kandler, O. 'The Amino Acid Sequence of the Peptide Moiety of Pseudomurein from *Methanobacterium thermoautotrophicum*', *Archives of Microbiology* (1979), *121*, 271–6

Millward, G.R. and Reaveley, D.A. 'Electron Microscope Observations on the Cell Walls of Some Gram-positive Bacteria', *Journal of Ultrastructure Research* (1974), *46*, 309–26

Oldmixon, E.H., Glauser, S. and Higgins, M.L. 'Two Proposed General Configurations for Bacterial Cell Wall Peptidoglycans Shown by Space-filling Molecular Models', *Biopolymers* (1974), *13*, 2037–60

Rogers, H.J. 'Biogenesis of the Wall in Bacterial Morphogenesis', *Advances in Microbial Physiology* (1979), *19*, 1–62

Rogers, H.J. *Bacterial Cell Structure* (Van Nostrand Reinhold, Wokingham, England, 1983)

Salton, M.R.J. and Shockman, G.R. *Beta Lactam Antibiotics: Mode of Action, New Developments and Future Prospects* (Academic Press Inc., New York, 1981)

Schleifer, K.H., Hammes, W.P. and Kandler, O. 'Effect of Endogenous and Exogenous Factors on the Primary Structures of Peptidoglycan', *Advances in Microbial Physiology* (1976), *13*, 246–92

Shockman, G.D. and Wicken, A.J. *Chemistry and Biological Activities of Bacterial Surface Amphiphiles* (Academic Press, London, 1982)

Shockman G.D. and Barrett J.F. 'Structure, function and assembly of cell walls of Gram-negative bacteria. *Annual Reviews of Microbiology* 37 (1983) 501–527

Stoddart, R.W. *The Biosynthesis of Polysaccharides* (Croom Helm, London/Macmillan, New York, 1984)

Tomasz, A. 'The Mechanism of the Irreversible Antimicrobial Effects of Penicillins: How Beta-lactam Antibiotics Kill and Lyse Bacteria', *Annual Review of Microbiology*, (1979), *33*, 113–37

Verwer, R.W.H., Nanninga, N., Keck, W. and Schwarz, U. 'Arrangement of Glycan Chains in the Sacculus of *E. coli*', *Journal of Bacteriology* (1978), *136*, 723–9

2 WALLS OF GRAM-POSITIVE BACTERIA

In Chapter 1 peptidoglycan was described as the most important component of bacterial cell walls, being vital for the normal functioning of the cell. Peptidoglycan accounts for approximately 50 per cent of the weight of the wall of Gram-positive bacteria. The remaining 50 per cent is made up from a variety of accessory polymers, the most important of which are teichoic and teichuronic acids. Proteins also play an important part in the structure of the wall in some Gram-positive organisms, but in general lipids are absent. Finally some Gram-positive organisms are encapsulated; that is to say they are enveloped by a layer of material, usually polysaccharide, which adheres to the wall but is not covalently attached.

The composition of Gram-positive bacterial walls has been determined by careful chemical analysis of walls which have been treated with a variety of agents to ensure that they are free from contaminating material. Walls are usually recovered from a suspension of cells broken in a device which subjects them to severe mechanical stress. Membrane fragments, ribosomes, and DNA all tend to stick to the walls in a broken cell suspension and it is obviously necessary to remove them before chemical analysis is carried out. Unfortunately the methods used to wash the walls have to be quite harsh to be effective. The two most widely used methods involve washing the walls in either hot aqueous phenol (45 per cent w/v at 90°C) or in a detergent such as sodium dodecyl sulphate (2 per cent w/v at 100°C). These procedures effectively remove much of the protein, lipid and nucleic acid contaminants from the walls, which are deposited from suspension by high-speed centrifugation and then extensively washed with saline and water to remove traces of phenol or detergent. These procedures yield 'clean' wall suspensions which appear white when freeze dried and are suitable for chemical analysis. Under the electron microscope they retain the shape of the cells from which they were obtained and they appear to be free of associated particles or material which might indicate contamination with cytoplasmic or membrane components (Figure 2.1). However, it would be unreasonable to suppose that these purified wall preparations accurately reflect the total wall structure surrounding intact growing cells. Although phenol or detergent extractions do not break covalent chemical bonds and will not release components which are covalently linked to the wall, they will strip off any molecules which are not covalently linked to the wall matrix. Therefore we must bear in mind that although analysis of pure walls may indicate that they contain say 50 per cent peptidoglycan and 50 per cent teichoic or teichuronic acids, other loosely associated components might also be important to the structure and

A

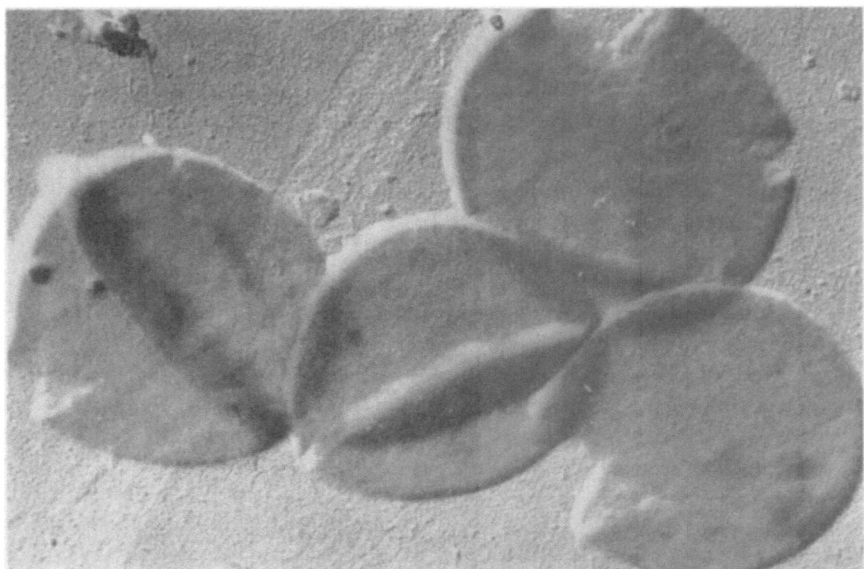

B

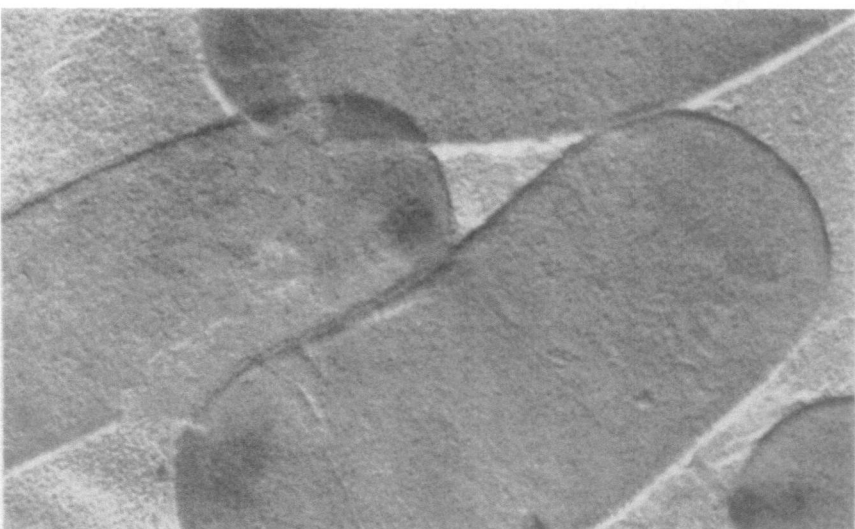

Figure 2.1: The Appearance of Isolated Gram-positive Bacterial Cell Walls under the Electron Microscope: A, *Staphylococcus aureus* (× 30,000); B, *Bacillus subtilis* (× 50,000). Whole cells were disrupted by vigorous shaking with glass beads, walls were deposited by cetrifugation, washed and treated with phenol to remove contaminating membrane and cytoplasmic material.

function of the walls of intact cells. We shall consider what these components might be and how they can be studied later in this chapter, but first we shall describe the major accessory wall polymers, teichoic and teichuronic acids.

Teichoic Acids

The name teichoic acid was originally chosen for a group of phosphate-containing polymers which could be isolated from the walls, membranes or capsules of virtually all Gram-positive bacteria. The first teichoic acids studied were linear polymers of either glycerol phosphate or ribitol phosphate in which the repeating units were joined by phosphodiester links. The name can also be applied to polymers in which sugar residues form an integral part of the chain of glycerol or ribitol phosphate units. Closely related polymers made up from sugar 1-phosphate units occur in the walls of some organisms and, whilst these are not strictly speaking teichoic acids, their properties and functions are closely related. Some typical teichoic acid structures are shown in Figure 2.2. All of the examples shown are polymers which are covalently linked to peptidoglycan and account for approximately 50 per cent of the weight of 'purified' walls. They can be released from the walls by a variety of means, the most commonly used being to stir wall suspensions with cold trichloroacetic acid for a long period (usually 24h with 10 per cent w/v trichloroacetic acid) or much more briefly with dilute alkali (for example for 1–2h with 0.1M-sodium hydroxide). Both methods hydrolyse the covalent linkage between teichoic acid and peptidoglycan (the precise nature of which will be discussed later) and release the teichoic acid as a soluble polymer, leaving the peptidoglycan as the insoluble remains of the wall. The size of the polymers released varies with the structures and the extraction conditions employed. Some breakdown of the teichoic acid chains by hydrolysis of the phosphodiester links is inevitable and chain lengths ranging from 5 to 35 units can be obtained. The chain length of the polymers in intact cells is thought to be of the order of 35 to 40 units. The high phosphate content means that the polymers are strongly acidic, they carry a net negative charge and are largely responsible for the negative surface charge carried by whole cells in which they occur. Attachment to the peptidoglycan framework of the wall is via a linkage unit containing a phosphodiester which couples one end of the teichoic acid chain to the 6-position of a muramic acid residue on the peptidoglycan.

Teichoic acid chains normally bear a number of substituents which have an important influence upon their molecular properties. For example, D-alanyl groups may be present, attached by an ester linkage between the carboxyl group of D-alanine and a free hydroxyl group on either glycerol

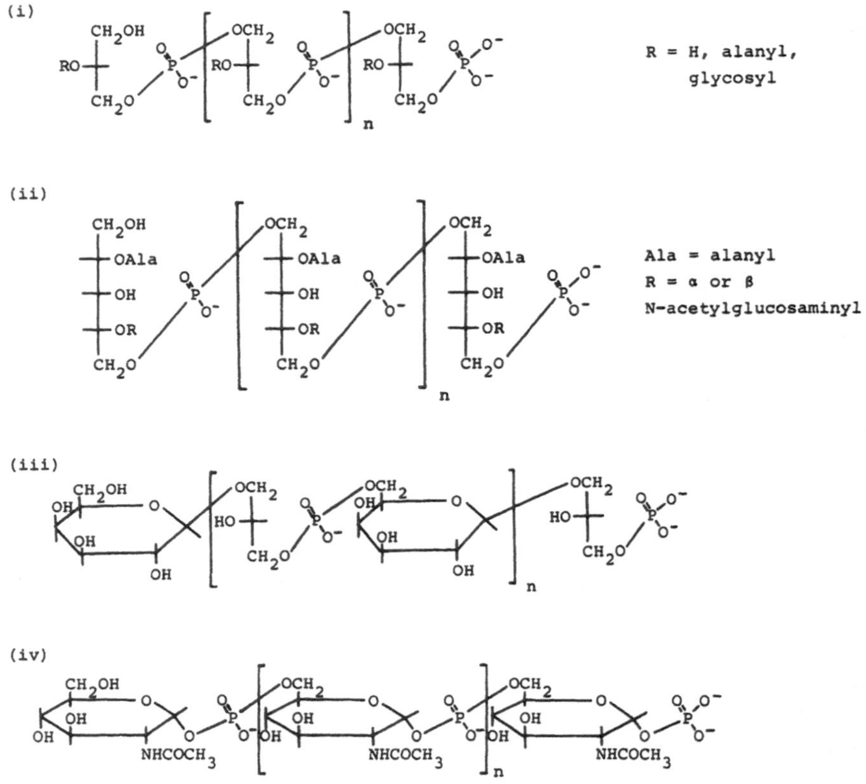

Figure 2.2: The Structure of Some Wall Teichoic Acids Comprising Repeating Units of:
(i), Glycerol Phosphate (1→3 linked); (ii), Ribitol Phosphate; (iii), Galactosylglycerol Phosphate;
(iv), *N*-acetylglucosamine Phosphate

or ribitol units of the teichoic acid chain. The free amino group from each alanine substituent is protonated under physiological pH conditions and effectively neutralises the negative charge on the adjacent phosphate group, so regulating the overall ionic nature of the polymer. The alanyl ester groups are unusually labile and the alanyl substituents are rapidly released from the teichoic acid chains by hydrolysis at pH values above 7. The alanyl ester content of teichoic acid in cell walls is clearly critically dependent upon the pH at which the cells are grown, and will appear to be very low if slightly alkaline conditions are used during extraction and purification. The reason why the alanine groups are of the D configuration rather than the more common L isomer is unknown, but it is interesting to recall that D-alanine is also found in peptidoglycan (Chapter 1).

Teichoic acids also frequently contain sugar substituents attached by α or β-glycosidic linkage to free hydroxyl groups on glycerol or ribitol. The function of the glycosyl substituents is unknown; they vary considerably in both nature and extent and do not appear to be essential to the cells although considerable metabolic energy is used to produce them.

Table 2.1 gives some idea of the distribution and diversity of teichoic acids among Gram-positive bacteria.

Teichuronic Acids

This name has been given to a group of wall polymers which have properties similar to the teichoic acids. They are linear polysaccharides containing uronic residues and are covalently attached at one end to peptidoglycan. Unlike teichoic acids, the repeating units of these polymers do not contain phosphate although they are probably linked to peptidoglycan by a single phosphodiester. Their negative charge is due to the carboxyl groups of the uronic acid residues. Some examples of teichuronic acid structures are shown in Figure 2.3. In general the polymers do not contain alanyl ester substituents. The distribution of teichuronic acids amongst Gram-positive bacteria is rather complex and in some cases is critically dependent upon the nutritional conditions under which the cells are grown.

The simplest cases are organisms such as *Micrococcus luteus* and *Bacillus megaterium* M46; walls of these organisms contain teichuronic acids regardless of the growth conditions. The organisms seem to be unable to produce teichoic acid. By contrast walls of *Staphylococcus aureus* nearly

Table 2.1: Distribution of Teichoic Acids Among Gram-positive Bacteria

Organism	Repeating unit of teichoic acid		
Staphylococcus aureus	ribitol phosphate		
Staphylococcus lactis 13 (*Micrococcus* sp. 13)	glycerol phosphate-*N*-acetylglucosamine phosphate		
Micrococcus sp.2102 (*M. varians*)	*N*-acetylglucosamine phosphate		
B. subtilis W23	ribitol phosphate		
B. subtilis 3610	glycerol phosphate		
B. licheniformis 9945	glucosylglycerol phosphate	}	separate
	galactosylglycerol phosphate	}	polymers
	glycerol phosphate		
B. stearothermophilus	glycerol phosphate (2, 3 linked)		
Lactobacillus buchneri	glycerol phosphate		
Streptococcus mutans	glycerol phosphate		

(i) $\left[\!\!\begin{array}{c}\end{array}\!\!\right.$4 Glucuronic acid 1→3 N-acetylgalactosamine 1$\left.\begin{array}{c}\end{array}\!\!\right]_{25}$

Bacillus licheniformis 6346
Bacillus subtilis W23 (phosphate-limited only)

(ii) $\left[\!\!\begin{array}{c}\end{array}\!\!\right.$4 Glucuronic acid 1→4 Glucuronic acid 1→3 N-acetylgalactosamine 1→6 N-acetylgalactosamine 1$\left.\begin{array}{c}\end{array}\!\!\right]_{n}$

Bacillus licheniformis 9945

(iii) $\left[\!\!\begin{array}{c}\end{array}\!\!\right.$4 N-acetylmannosaminuronic acid 1→6 Glucose 1$\left.\begin{array}{c}\end{array}\!\!\right]_{40}$

Micrococcus luteus

Figure 2.3: The Structure of Some Teichuronic Acids

always contain teichoic acid, most strains being incapable of producing teichuronic acid. *Bacillus subtilis* and *Bacillus licheniformis* have the ability to produce teichoic and teichuronic acids. When growing in a medium containing excess of phosphate cells of *B. subtilis* var. niger produce walls containing a polyglycerol phosphate teichoic acid but no teichuronic acid. However, when the amount of phosphate is reduced so that it becomes the growth-limiting nutrient the cells cease production of teichoic acid and, instead, incorporate a teichuronic acid into the walls. Similar behaviour occurs with *B. subtilis* W23 which contains a polyribitol phosphate teichoic acid when excess of phosphate is available and a teichuronic acid when phosphate is scarce. Strains of *B. licheniformis* acts in a similar way although a teichuronic acid is produced together with the teichoic acid under conditions of excess phosphate. It is not only the availability of phosphate which controls the synthesis of teichoic and teichuronic acids. The concentration of magnesium and sodium are important factors, as is the growth rate of the cells. By studying the changes in cell wall composition which occur as a result of altering the growth conditions it is possible to draw some conclusions as to the functional importance of teichoic and teichuronic acids to the cells. We shall see later that one important function of teichoic acid is to maintain an adequate supply of magnesium for the cells.

Linkage of Teichoic and Teichuronic Acids to the Peptidoglycan Layer

Teichoic and teichuronic acids can only be separated from peptidoglycan in the cell wall by treatment with reagents, usually dilute acid or alkali, which hydrolyse covalent chemical bonds. No amount of washing with salt solutions or detergents will effect their release. Careful chemical analysis has revealed that teichoic acids are coupled to peptidoglycan by a special linkage unit, which is a short chain made up of three glycerol molecules and one molecule of *N*-acetylglucosamine each joined by phosphodiesters. The glycerol end of the linkage unit is joined through a phosphodiester to the main teichoic acid chain and the *N*-acetylglucosamine is joined by a phosphodiester to the OH group on position 6 of a muramic acid residue on peptidoglycan. Between 5 and 10 per cent of the muramic acid residues in peptidoglycan have teichoic acid chains linked to them. The linkages on positions 1 and 4 of *N*-acetylglucosamine are particularly susceptible to acid and alkaline hydrolysis respectively (Figure 2.4) which explains why teichoic acids can be released from the wall without extensive degradation of the polymer chains.

The linkage unit appears to be a common feature in walls containing teichoic acids, regardless of the composition of the teichoic acid. For example, in *S. aureus* and strains of *B. subtilis* containing ribitol teichoic acids the triglycerolphosphate-*N*-acetylglucosamine linkage unit is clearly distinguishable from the main polymer chain. Similarly, in organisms containing sugar-1-phosphate polymers such as *Mircrococcus* sp2102, which has a polymer of repeating units of *N*-acetylglucosamine linked by phosphodiesters at the 1 and 6 positions, the triglycerolphosphate-*N*-acetylglucosamine linkage unit has been shown to link the polymer to the peptidoglycan. In organisms containing polyglycerol phosphate teichoic acids the triglycerol phosphate part of the linkage unit is indistinguishable from the main chain but the polymer is still linked to peptidoglycan in the same way, i.e. by a phosphodiester between muramic acid and *N*-acetylglucosamine on the end of the teichoic acid chain.

At the present time there is no explanation for the occurrence of a common linkage unit among bacteria which have chemically diverse teichoic acids. The unit is synthesised separately and assembled in a different manner from the main teichoic acid chain; perhaps in this way the organisms are able to regulate and co-ordinate synthesis of teichoic acid with that of peptidoglycan so that during growth the wall is expanded in a controlled, orderly manner.

In contrast to the teichoic acids, little is known yet about the way in which teichuronic acids are linked to peptidoglycan. It appears that no special linkage unit is involved; in some cases the end of the teichuronic acid chain is linked to the 6 position of muramic acid in the peptidoglycan by a phosphodiester group. The occurrence of a single phosphate group in the teichuronic acid/peptidoglycan structure is quite surprising since no other phosphates are involved in either of the polymers. Presumably insertion of the phosphate group to link the polymers is an important step in the biosynthetic sequence by which control over wall assembly is exerted.

Other Polymers Associated with the Gram-positive Cell Wall

The walls of virtually all Gram-positive bacteria that have been studied have been shown to contain either teichoic or teichuronic acids, whilst some contain a mixture of both polymers. Any other wall components that are not tightly bound will be released under the harsh conditions used to purify cell walls for analysis. It is quite possible that many loosely bound wall components remain unrecognised. The peptidoglycan/teichoic or teichuronic acid structure should be thought of as the basic framework or skeleton of the wall. It is undoubtedly responsible for the shape, strength and bulk of the wall, but associated with it, or protruding through it from the cytoplasmic membrane are a number of macromolecules which are also of vital importance to the cells. Surface structures such as pili and flagella are also associated with the wall and membrane, these will be considered in Chapter 4.

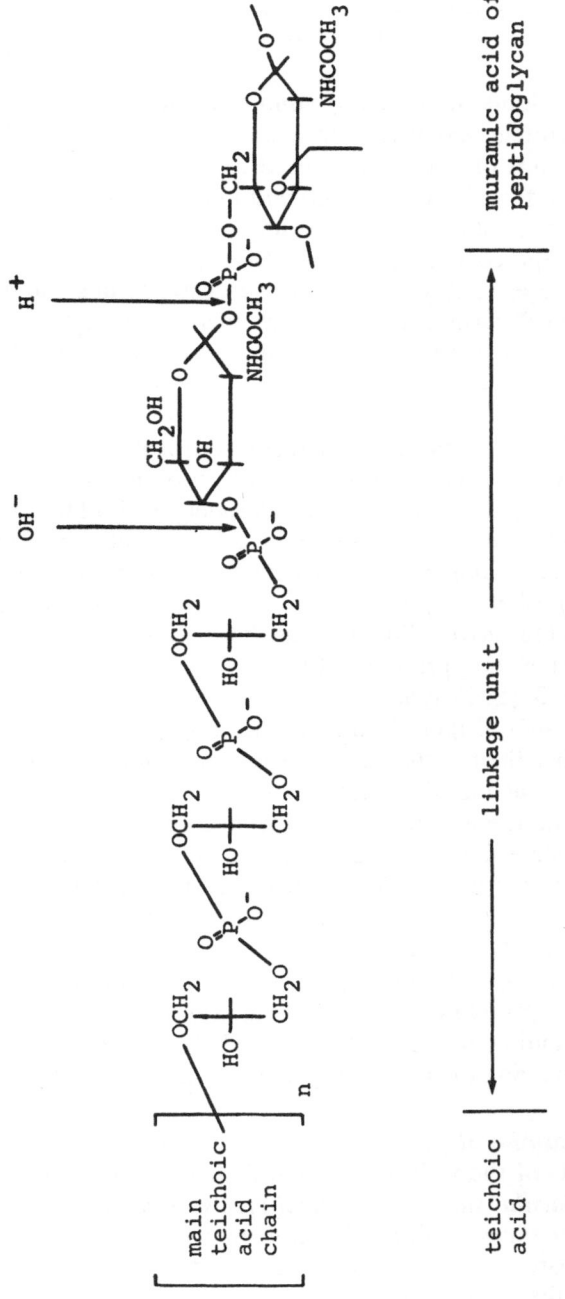

Figure 2.4: The Linkage Unit Which Joins Teichoic Acid to Peptidoglycan Showing Sites Sensitive to Hydrolysis by Dilute Acid or Alkali

Proteins

A variety of proteins are now recognised as wall components, some are structural, others are enzymes involved in wall assembly or modification.

M Protein. Some streptococci contain an important surface component called M protein which is tightly associated with the wall. Its molecular weight depends on the method used to release it from the wall; reported values range from 32,000 to 180,000 implying that it forms aggregates of subunits in the wall. M Protein is important because it enhances the virulence of streptococci by protecting them from destruction in the body by white cells. Presumably the protein aggregates form some sort of protective coat on the cell surface which protects them from engulfment and digestion by the phagocytes. At present it is not known how M protein is linked to other wall components.

Protein A. One of the best characterised proteins found in the walls of Gram-positive bacteria is protein A, which occurs in certain strains of *Staphylococcus aureus*. It has a molecular weight of 42,000 and is covalently linked by its carboxy terminus to an unknown site on peptidoglycan. The protein has an elongated conformation, part is embedded in the cell wall whilst the rest of the polypeptide, consisting of four similar regions, protrudes into the surrounding medium. The protruding portions of protein A have the interesting property of binding antibody molecules of the immunoglobulin G (IgG) type. Each molecule of protein A can bind two or more molecules of IgG. Binding occurs at the constant region (Fc portion) of IgG rather than at the variable site responsible for the specific antigen-antibody interaction. Consequently protein A is a valuable tool for immunologists who use it to recover IgG from solution in various ways: either in its natural state linked to whole cells of *S. aureus,* which can be conveniently removed from suspension by centrifugation after binding IgG, or in affinity chromatography where purified protein A is linked covalently to an insoluble support such as agarose.

The reason why *S. aureus* produces protein A remains something of an enigma. The protein does also occur in a free form which is released from the cells. Binding of IgG by the extracellular protein A might have survival value by diverting the opsonising antibody from the cells.

Proteins Forming Regular Surface Arrays. Proteins have been detected as components of the walls of many rod-shaped Gram-positive bacteria, for example *Bacillus subtilis, B. licheniformis, B. polymyxa, B. sphaericus* and *Clostridium thermosaccharolyticum.* In some cases the proteins form a regularly patterned layer on the cell surface that can be clearly distinguished in the electron microscope by negative staining. Usually the layers can be removed by mild treatment such as washing with a high concentration (e.g. 8 M) of an aqueous solution of urea, indicating that the protein is not covalently linked to the wall. When removed from the cells the proteins retain the ability to associate together forming regularly patterned sheets; under appropriate conditions they can be reassembled on the surface of cells. Regular arrays of proteins on the cell surface are by no means common to all Gram-positive bacteria, but when they do occur they are major wall components and are presumably important determinants of the surface properties of the cells.

Autolytic Enzymes. Proteins responsible for modifying the wall structure, by breaking covalent links in the peptidoglycan at various specific points, are present in the walls of most Gram-positive bacteria. Whilst at first sight enzymes which damage the peptidoglycan might be thought of as potentially lethal to the cells, it is obvious that the walls must expand during growth and undergo limited breakdown at the site of cell division. This is the role of the autolytic enzymes; their presence is clearly demonstrated when cells from an actively growing broth culture are harvested. If the cells are left for even a short time as a pellet in a centrifuge tube or are resuspended in warm buffer they start to lyse rapidly. Autolytic activity can only be held in check by harvesting and maintaining the cells at a low temperature (0–5°C). In some cases autolytic activity is so great that it can be exploited to prepare protoplasts by removing the wall completely in an osmotically supporting medium. For example, when exponential phase cells of *Streptococcus faecalis* are suspended in 0.04M-ammonium acetate buffer, pH6.7, containing 1mM-magnesium acetate and 0.5M-sucrose the walls are completely digested by autolytic activity within 3 to 6h at 37°C. Because of the high sucrose content of the medium the osmotically fragile protoplasts that are formed (usually termed autoplasts) remain intact for up to 24h.

Autolytic enzymes usually remain associated with the walls when cells are broken and centrifuged to remove the cytoplasmic contents. Even after several washes in saline autolytic activity can be demonstrated when the walls are suspended in a buffer under the optimum conditions for autolytic enzyme activity. In fact the activity can only be eliminated by boiling the walls or by extracting them with phenol or detergent solutions. Clearly such potentially lethal activity contained within the walls must be carefully controlled if the cells are to survive, grow and divide in an ordered way. The control of the autolysins appears to be exerted at least in part by the accessory charged polymers associated with the peptidoglycan, i.e. the teichoic and/or teichuronic acids. The most clear cut demonstration is provided by the organism *Streptococcus pneumoniae.* The wall teichoic acid is a complicated structure containing choline phosphate substituents (Figure 2.5). The organism is unable to synthesise choline, it therefore has

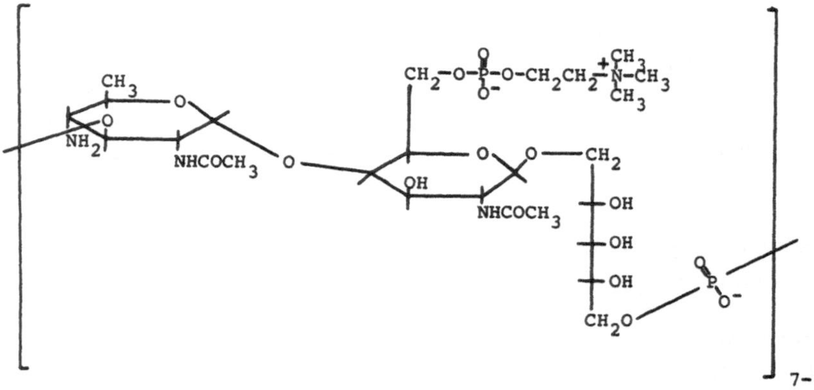

Figure 2.5: The Wall Teichoic Acid of *Streptococcus pneumoniae*

to obtain choline from the growth medium and incorporate it into the wall teichoic acid. Normal cells grown in this way have an active autolytic enzyme system, the major system being an amidase which breaks the linkage between muramic acid and L-alanine in the peptidoglycan. The cells grow characteristically as pairs of cocci (hence the original name of *Diplococcus pneumoniae* and are prone to spontaneous lysis when harvested from the growth medium. When ethanolamine is supplied in the growth medium in place of choline it is incorporated into the wall teichoic acid as ethanolamine phosphate in place of choline phosphate. This simple change in the nature of the wall teichoic acid has a dramatic effect on the autolytic activity of the wall. In effect the autolysins are inhibited, the cells no longer lyse spontaneously and, instead of growing in pairs, they grow as long chains of cocci, the cells being unable to separate after division.

Autolytic enzymes bind tightly to teichoic and teichuronic acids in cell walls and it is thought that the binding regulates the activity. In the case of *Streptococcus pneumoniae* the autolysin is presumably permanently inhibited when bound to the teichoic acid containing ethanolamine. It is difficult to envisage how enzyme activity is switched on and off during growth, especially since the wall is outside the direct metabolic influence of the systems in the cytoplasm and the cytoplasmic membrane. One possible link is the occurrence of membrane teichoic acids. These are exclusively polymers of polyglycerol phosphate linked at one end to a lipid molecule which anchors them to the outer face of the cytoplasmic membrane. The teichoic acid chains protrude from the membrane into the matrix of the wall and possibly beyond into the surrounding medium. Like the wall teichoic acids, these polymers also bind autolytic enzymes in the wall. Control of autolytic activity might be achieved by a combination of wall and membrane teichoic acids. The possible interaction of wall and membrane teichoic acids will be discussed later in the chapter.

Neutral Polysaccharides

Whereas teichuronic acids are essentially polysaccharides containing free carboxyl groups, neutral polysaccharides are also found in the walls of some Gram-positive bacteria. The walls of certain streptococci contain uncharged polysaccharides which appear to be covalently linked to the peptidoglycan through a phosphodiester to position 6 of muramic acid in the same way as teichoic and teichuronic acids. Group A streptococci contain a neutral polysaccharide made up from L-rhamnose and *N*-acetyl-glucosamine units, whereas in group C streptococci the polysaccharide contains *N*-acetylgalactosamine in place of *N*-acetylglucosamine. The allocation of letters A and C to these polysaccharides arises from the way in which haemolytic streptococci are grouped serologically. These neutral polysaccharides are antigenic and form the basis of the serological grouping designated by Lancefield A to O. Recognition of their existence as antigens was made long before their composition was determined. It seems that the chemical basis for the serological difference between the groups resides in the aminosugar, e.g. *N*-acetylglucosamine in group A and *N*-acetylgalactosamine in group C.

Membrane Teichoic Acids and Capsules

So far we have discussed the major components which are linked together to form the Gram-positive cell wall. There are a number of additional components which need now be considered; although they are not covalently attached to the wall they are nevertheless intimately associated with it and their presence has a significant effect upon the properties of the wall. The first of these components is called lipoteichoic acid or membrane teichoic acid. It is a polyglycerol phosphate teichoic acid covalently linked to a glycolipid molecule, which is presumed to anchor the molecule to the outer face of the cytoplasmic membrane. The polyglycerol phosphate chain is 25 to 30 units long and protrudes through the matrix of the cell wall so that it can be detected on the outer surface of the cells. The second group of accessory wall components are capsules. These are usually large poly-saccharide molecules which envelop the walls and are loosely associated with the outer surface of the cell wall. One property of the capsule which can be important in infections is the protection it gives the cells against attack by phagocytic leukocytes.

Membrane of Lipoteichoic Acids

All Gram-positive bacteria that have a teichoic acid as part of their cell wall have, in addition, a lipoteichoic acid which can be found associated with the cytoplasmic membrane (hence the alternative name of membrane teichoic acid). Unlike the wall polymers, lipoteichoic acids are exclusively of the polyglycerol phosphate type in which glycerol units are linked in a 1,3 manner by phosphodiesters (Figure 2.6). Therefore an organism such as *Staphylococcus aureus* which has a ribitol teichoic acid in its wall has a glycerol lipoteichoic acid associated with its cytoplasmic membrane. The glycerol units in the polymer chain are substituted with glycosyl and D-alanine esters to a variable extent. The alanyl substituents have a considerable influence upon the ionic properties of the polymer and, possibly, upon the extent to which it is able to participate in the biosynthesis and assembly of the wall, as will be described later. The major difference between wall and membrane teichoic acids is the occurrence of the glycolipid moiety, covalently linked to one end of the polymer chain of the latter. The glycolipid consists of a diglyceride unit (glycerol with two long-chain fatty acids ester-linked to it) and a number of sugars. The nature of the sugar units varies from organism to organsim. In *Staphylococcus aureus* the glycolipid contains two glucose units 1→6 linked to form the disaccharide, gentiobiose, which couples the diglyceride unit to the polyglycerol phosphate chain. The complete molecule has amphipathic properties due to the presence of the hydrophobic glycolipid and the hydrophilic teichoic acid chain. It is generally assumed that the glycolipid anchors the lipoteichoic acid to the outer face of the cytoplasmic membrane, with the fatty acids embedded in the outer leaflet of the membrane. Magnesium ions are important in maintaining the association with the membrane; in their absence, lipoteichoic acid is released from membranes but, in the presence of concentrations of 10mM or more, it remains with the membranes. Free glycolipid also occurs in the cytoplasmic membrane and the lipoteichoic acid contains the glycolipid which is characteristically found in the organism. It seems reasonable therefore to describe membrane teichoic acids as being covalently linked to the cytoplasmic membrane via the glycolipid component (whereas the wall teichoic acid is covalently linked to the wall via the linkage unit to the muramic acid of the peptidoglycan).

The reason for discussing the lipoteichoic acids here is that in intact cells the polyglycerol phosphate chain extends through the wall in such a way that it can be recognised on the cell surface as an antigen. It is therefore effectively a wall component and gives the walls a number of important properties. For example lipoteichoic acids, like the wall teichoic acids, bind magnesium ions tightly and are thought to help the organism maintain an adequate supply from the surrounding medium. One attractive idea is that

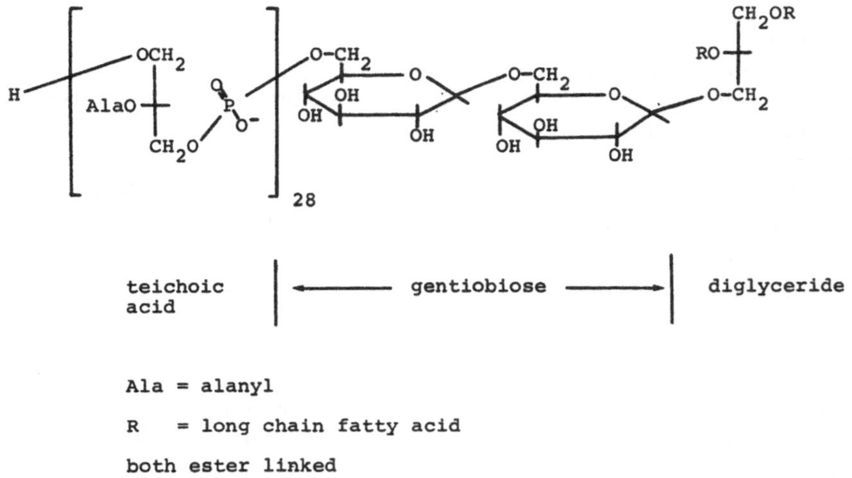

Figure 2.6: The Structure of Lipoteichoic Acid of *Staphylococcus aureus*

the two forms of teichoic acid combine to channel magnesium ions from the outer surface of the cells to the cytoplasmic membrane. Interestingly, some of the membrane-bound enzymes which require magnesium to function preferentially utilise magnesium which is bound to teichoic acids rather than free magnesium in solution. Possibly these enzymes accept magnesium ions in a partially dehydrated form by direct transfer from the teichoic acids without intermediate solvation of the metal ions. Lipo-teichoic acids are probably more important to the cells than the wall teichoic acids since when grown under conditions where the supply of phosphorus is limited, the cells replace the phosphate-rich wall teichoic acid with teichuronic acid but continue to produce some lipoteichoic acid.

A second important property of lipoteichoic acids is their ability to bind to autolytic enzymes and inhibit their action. In intact cells it is probably the wall teichoic acids which control autolytic enzymes present in the wall and there is considerable specificity in the interaction between different autolysins and teichoic acid structures. Lipoteichoic acids appear to inhibit autolysins in a non-specific manner which is related to their acidic, amphi-pathic nature (acidic phospholipids such as diphosphatidylglycerol behave in a similar way). During growth and division some lipoteichoic acid is excreted into the medium and this free, extracellular material probably protects the cells against the action of external lytic enzymes.

A third property concerns the adhesion of bacteria to surfaces. Lipoteichoic acid is one of a number of wall-associated polymers (including polysaccharides and proteins) which can play a part in assisting the cells to adhere to surfaces. In the case of *Streptococcus mutans* lipoteichoic acid, wall proteins and polysaccharides are all involved in sticking the cells to the surface of teeth. Acids produced by the cells growing on the teeth are responsible for demineralisation of the tooth surface and the formation of cavities.

A few organisms do not produce teichoic acid in their walls or membranes under any conditions. For example, the walls of *Micrococcus luteus* contain a teichuronic acid (Figure 2.3) and the membrane contains a lipomannan polymer which has properties which are closely related to those of lipoteichoic acids. The lipomannan contains a chain of about 60 mannose units linked at one end (the reducing end) to a diglyceride (a glycerol molecule bearing two ester-linked fatty acid substituents). The diglyceride anchors the lipomannan to the cytoplasmic membrane in the same way that the glycolipid anchors the lipoteichoic acids. The mannan chain bears a number of succinic acid residues which give the polymer a negative charge and enable it to bind magnesium ions efficiently. The lipomannan of *M. luteus* is therefore presumed to function in the same way as the lipoteichoic acids of other Gram-positive bacteria.

Capsules

Capsules associated with Gram-positive bacteria are usually polysaccharides, although other polymers occur in certain cases. The pneumococci (*Streptococcus pneumoniae*) possess polysaccharide capsules which are of vital importance to the pathogenic potential of the organism. If the capsule is lost the organisms become sensitive to killing by phagocytes and are no longer virulent. Over 85 immunologically distinct capsular polysaccharides have been distinguished and are used as a basis for typing of strains. Of these, types 1 to 8 and 18 are responsible for 80 per cent of the cases of pneumococcal pneumonia and over 50 per cent of the cases of pneumococcal bacteraemia in adults. In children the most frequent cases of infection are types 6, 14, 19 and 23. Structures have been determined for many of the capsular types, they range from relatively simple polysaccharides to complex molecules, some of which are reminiscent of the teichoic acids since they contain ribitol phosphate or glycerol phosphate units as part of their repeating units. The structures of three such capsular polysaccharides from *Streptococcus pneumoniae* types 3, 34 and 11A are shown in Figure 2.7. It must be emphasised that the capsular polysaccharides (even the teichoic acid-like polymers) are not covalently linked to the wall, although they are closely associated with it and remain associated with the cells when they are harvested and washed. All of the pneumococci contain a genuine

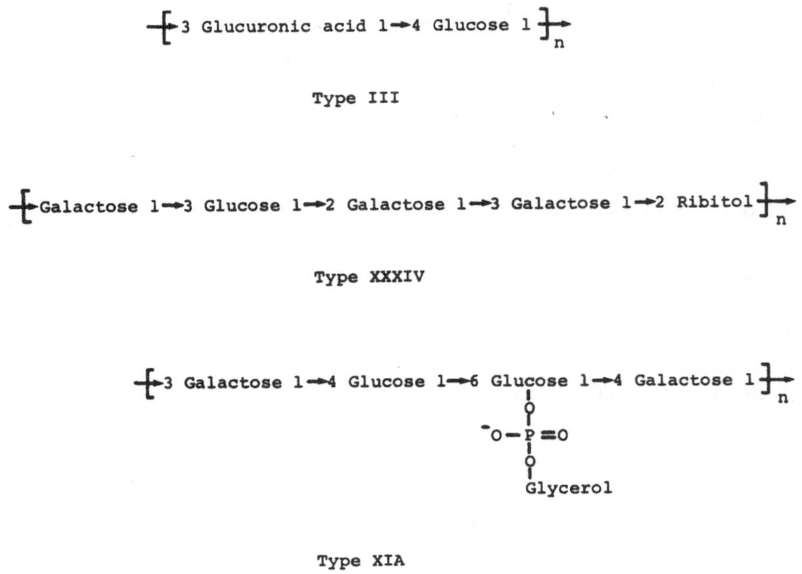

$$\text{--}\!\!\left[\text{-3 Glucuronic acid 1}\!\!\rightarrow\!\!\text{4 Glucose 1}\right]_n\!\!\text{--}$$

Type III

$$\text{--}\!\!\left[\text{-Galactose 1}\!\!\rightarrow\!\!\text{3 Glucose 1}\!\!\rightarrow\!\!\text{2 Galactose 1}\!\!\rightarrow\!\!\text{3 Galactose 1}\!\!\rightarrow\!\!\text{2 Ribitol}\right]_n\!\!\text{--}$$

Type XXXIV

$$\text{--}\!\!\left[\text{-3 Galactose 1}\!\!\rightarrow\!\!\text{4 Glucose 1}\!\!\rightarrow\!\!\text{6 Glucose 1}\!\!\rightarrow\!\!\text{4 Galactose 1}\right]_n\!\!\text{--}$$

$$\begin{array}{c} O \\ \| \\ {}^-O\text{---}P=O \\ \| \\ O \\ | \\ \text{Glycerol} \end{array}$$

Type XIA

Figure 2.7: The Structure of Some Capsular Polysaccharides of *Streptococcus pneumoniae*

wall teichoic acid which is also antigenic and was originally termed C-substance. Its structure appears to be identical for all the different capsular types.

Although most strains of *Staphylococcus aureus* are not encapsulated a few strains have been isolated which possess antiphagocytic capsules. The structures have not been fully established but they appear to be made up from repeating units of aminosugars. The capsules are antigenic and can be used to distinguish between encapsulated strains immunologically.

All of the Gram-positive capsular polysaccharides so far described are heteropolysaccharides, that is to say their repeating units are made up from more than one sugar residue. Their shape and solution properties depend upon the linkage of the components of the repeating unit and the extent to which the chain is branched. Some of the capsules have extremely complex structures and many remain to be established. By contrast, homopolysaccharide capsules are relatively simple in structure, being composed of repeating units of a single sugar. Examples are the dextrans, polymers made up from repeating glucose units produced by organisms such as *Leuconostoc mesenteroides* and *Streptococcus mutans*. The dextrans from *L. mesenteroides* have been used medically as blood plasma substitutes. The adherence of *Streptococcus mutans* to teeth and the resulting production of cavities is due in part to the dextran produced by the organism. Two forms of dextran are produced: a 'water insoluble' form which remains with the cells as a capsule, and a 'water soluble' form which is released as extracellular slime. Both forms are important in the formation of dental plaque. Another group of homopolysaccharides are the levans, polymers of fructose which are produced by *Streptococcus salivarius* and some bacilli. Finally the organism responsible for anthrax, *Bacillus anthracis* produces an unusual capsule which is a polypeptide, composed entirely of D-glutamic acid units.

In addition to helping organisms adhere to surfaces (e.g. *Streptococcus mutans* and teeth) and protecting them from body defence mechanisms, such as the bactericidal action of serum or ingestion and killing by phagocytes (e.g. the pneumococci and *Staphylococcus aureus*) capsules probably protect cells from desiccation in natural environments which are prone to periodic drying. Their survival value to the cells therefore justifies the metabolic effort required for their production.

Biosynthesis and Assembly of the Gram-positive wall

A great deal of attention has been given to the mechanisms by which peptidoglycan is synthesised and assembled. This is partly because of its fundamental importance to bacteria and partly because extensive research has been carried out into the mode of action of antibiotics which inhibit its synthesis. The teichoic acids, teichuronic acids, neutral and acidic polysaccharides, and proteins of Gram-positive walls have not been investigated on quite the same scale. Nevertheless, in some cases an insight into the processes involved is now being gained at the molecular level. The picture that is beginning to emerge is both complex and fascinating. The bacterium is faced with the problem of expanding its wall before division takes place. The expansion must occur in an ordered and controlled manner so that the cell retains both its mechanical integrity and its characteristic shape. Synthesis and assembly of the different wall polymers must be coordinated so that the components are brought together at the correct time and in the right place. The evident ease with which bacteria accomplish this task must surely inspire microbiologists to continue to probe the mechanisms. One way in which synthesis of different polymers might be controlled and coordinated is by sharing common precursors, enzymes or intermediates. One possible common factor linking the synthesis of several components is the polyisoprenoid alcohol phosphate lipid intermediate, undecaprenyl phosphate (Figure 1.11). This important molecule is located exclusively in the cytoplasmic membrane. It acts as the common initial acceptor for the precursors of peptidoglycan, teichuronic acid, most of the capsular polysaccharides and the teichoic acid linkage unit. The precursors of the polymers, which are synthesised first as nucleotide derivatives in the cytoplasm are transferred to the lipid on the inner face of the cytoplasmic membrane with release of the nucleotide. The lipid then carries the precursors across the membrane to the outer face where they are incorporated into polymers.

Teichoic Acids

Most wall polymers were discovered and characterised as a direct result of the chemical analysis of wall fractions whereas the details of their biosynthesis followed much later. Teichoic acids are an interesting exception to this general rule: their discovery as a group of polymers was made after the detection of trace amounts of their precursors: cytidine diphosphate ribitol (CDP-ribitol) and cytidine diphosphate glycerol (CDP-glycerol). The structures of these precursors are shown in Figure 2.8. Then, by analogy with the function of sugar nucleotides in the biosynthesis of polysaccharides, it was thought that the cytidine compounds might be required for the synthesis of polymers containing glycerol phosphate and ribitol phosphate residues. Careful analysis of Gram-positive bacteria led to the discovery of large amounts of such polymers in the walls and smaller amounts in the cell contents fractions of disrupted cells, later shown to be lipoteichoic acid.

The nucleotide precursors of teichoic acids are synthesised in the cytoplasm but they are assembled at some point in the cytoplasmic membrane and coupled to peptidoglycan at a growing point in the cell wall. The acceptor molecule in the cytoplasmic membrane which acts as the intermediate site for assembly and extension of the teichoic acid chain was

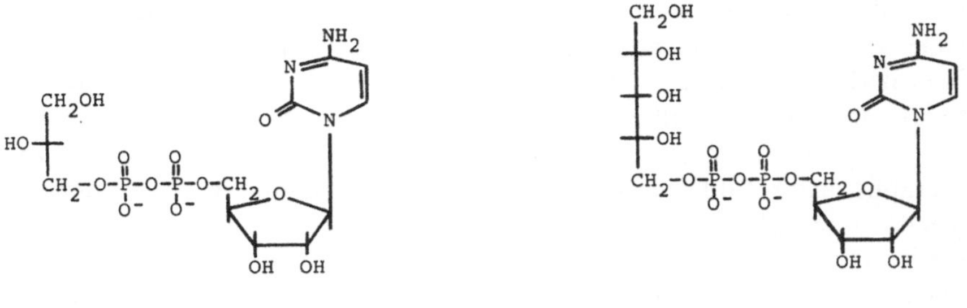

CDP-glycerol   CDP-ribitol

Figure 2.8: The Structures of the Teichoic Acid Precursors, CDP-glycerol and CDP-ribitol

originally thought to be the same undecaprenyl phosphate (lipid intermediate) used for peptidoglycan assembly. However, the situation appears
to be a little more complicated. The linkage unit, composed of three
glycerol phosphate and one *N*-acetylglucosamine residues is indeed
assembled on the undecaprenyl phosphate acceptor using CDP-glycerol
and UDP-*N*-acetylglucosamine as precursors. However, the mechanism by
which the main teichoic acid chain is assembled is controversial and has not
been firmly established. Studies with cell-free systems containing the
enzymes and precursors required for teichoic acid synthesis have shown
that a molecule termed lipoteichoic acid carrier (LTC) can act as the
acceptor on which a chain of 30 to 40 glycerol phosphate or ribitol phosphate units are assembled. Some confusion has arisen over the precise
nature of LTC, many workers have found it elusive and difficult to
characterise. Its structure appears to be similar to that of the membrane
teichoic acids, the best acceptor activity being displayed by molecules
containing about 20 glycerol phosphate units in a chain terminating in a
glycolipid (as shown for lipoteichoic acid in Figure 2.6) but lacking ester-
linked D-alanine substituents. Whilst LTC molecules have quite clearly
been shown to function in cell-free systems, it seems unlikely that they are
the true acceptors on which teichoic acids are assembled in whole cells.
LTC molecules probably function in cell-free systems because they are
analogous in structure to the true physiological receptor, which is most
likely the linkage unit. Linkage unit is synthesised before the teichoic acid
chain and is therefore available in the cytoplasmic membrane, attached to
undecaprenyl phosphate. The main chain, consisting of up to 40 ribitol
phosphate or glycerol phosphate units is then assembled on the linkage
unit/undecaprenyl phosphate, either by a single transfer of a preformed
chain from LTC (which is doubtful but certainly occurs in cell-free
systems) or by sequential additions of 40 individual molecules from the
nucleotide precursors CDP-ribitol or CDP-glycerol. When the complete
chain of teichoic acid and linkage unit has been assembled on the lipid
intermediate it is transferred to the peptidoglycan and covalently attached
to it by formation of a phosphodiester link between the *N*-acetylglucosamine
on the end of the linkage unit and the free OH group on position 6 of a
muramic acid residue on the peptidoglycan. An outline of the sequence of
events in *Staphylococcus aureus* is shown in Figure 2.9. As suggested
earlier, the involvement of the same lipid intermediate, undecaprenyl
phosphate, in both teichoic acid and peptidoglycan synthesis might provide
the point of control and coordination over wall polymer synthesis and
assembly. It should also be remembered that whereas the chemical nature
of the teichoic acid chain varies considerably amongst different organisms
(Table 2.1, Figure 2.2) the linkage unit always appears to be the same,
implying a fundamental importance. The alanyl and glycosyl substituents
are probably added during assembly of the polymers rather than to existing

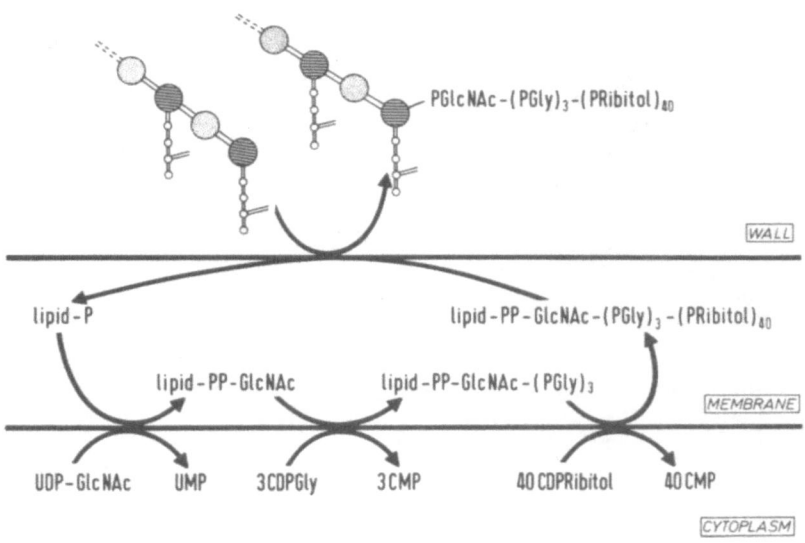

Figure 2.9: Synthesis of Teichoic Acid and Attachment to Peptidoglycan in *S. aureus*

unsubstituted chains. Enzyme systems capable of transferring the substituents have been isolated: glycosyl substituents (e.g. glucosyl or *N*-acetylglucosaminyl groups) are added to teichoic acids from the respective UDP-sugar precursors and alanyl substituents are added by ligase enzymes which utilise D-alanine and ATP.

Teichuronic Acids

Comparatively little is known about the mechanism by which teichuronic acids are synthesised. Sugar nucleotides act as the precursors: the alternating units of *N*-acetylgalactosamine and D-glucuronic acid found in the teichuronic acids of *B. licheniformis* 6346 and *B. subtilis* W23 (grown under phosphate limitation) are derived from UDP-*N*-acetylgalactosamine and UDP-glucuronic acid. In *M. luteus* UDP-*N*-acetylmannosaminuronic acid and UDP-glucose are the precursors of the teichuronic acid which is a linear polymer of repeating *N*-acetylmannosaminuronic acid and D-glucose units. One possible sequence of steps leading to a complete teichuronic acid chain linked to peptidoglycan in *M. luteus* is shown in Figure 2.10; in many respects it is analogous to the scheme for teichoic acid

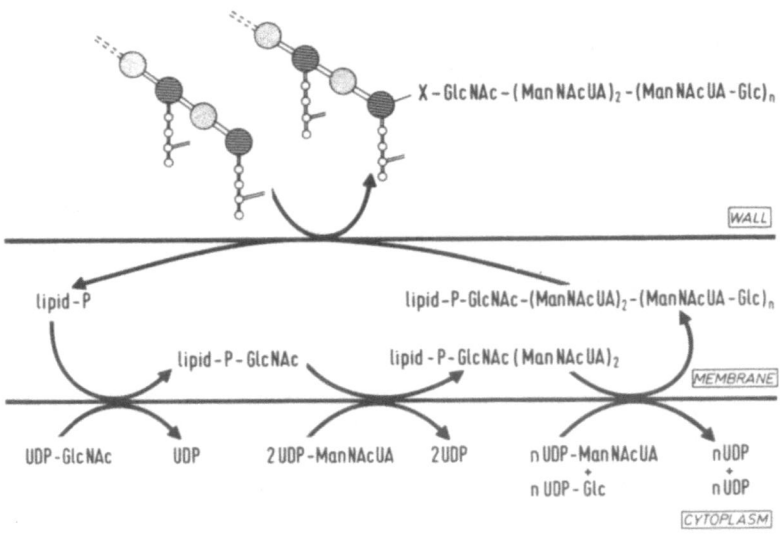

Figure 2.10: Synthesis of Teichuronic Acid and Attachment to Peptidoglycan in *Micrococcus luteus*

synthesis shown in Figure 2.9. There appears to be three stages in the process: first, a linkage piece comprising one unit of *N*-acetylglucosamine and two *N*-acetylmannosamin uronic acid units is assembled on a lipid intermediate carrier (presumably undecaprenyl phosphate). In the second stage the main teichuronic acid chain is assembled on the linkage unit/lipid intermediate molecule by transfer of alternating glucose and *N*-acetyl-mannosaminuronic acid units from the respective nucleotide precursors. Finally the main teichuronic chain and linkage unit is transferred from the lipid carrier to peptidoglycan. The linkage is probably formed at position 6 of muramic acid by a phosphodiester group. It must be emphasised that this biosynthetic sequence might not apply to all teichuronic acids and indeed the precise chemical nature of the units linking teichuronic acids with peptidoglycan have not yet been established in the same detail as the teichoic acid linkage units.

Lipoteichoic Acids

The polyglycerol phosphate chain of lipoteichoic acid is not synthesised from CDP-glycerol in the same way as the wall teichoic acids. The glycerol phosphate units are of the opposite stereochemical series to those found in the wall teichoic acids and are derived from the phospholipid phosphatidyl-glycerol (p. 107). The units are polymerised and linked to a glycolipid which is characteristic of the species and is usually also present in the organism as an unsubtituted membrane glycolipid. Alanyl substituents are added to the lipoteichoic acid by a ligase in the same way as for the wall teichoic acids.

Capsular Polysaccharides

The capsular polysaccharides of *Streptococcus pneumoniae* are synthesised in a similar way to the teichoic and teichuronic acids using sugar nucleotide precursors and the undecaprenyl phosphate lipid intermediate. The sugar nucleotides used for the synthesis of the simplest polymer (type III, Figure 2.7) are UDP-glucose and UDP-glucuronic acid. The enzymes responsible for assembling the sugars on the lipid intermediate are located in the cytoplasmic membrane but the means by which the growing polysaccharide is channelled through the wall to take up its position on the outermost surface of the cell is unknown.

Not all capsular polysaccharides are synthesised by assembly on a lipid carrier from nucleotide precursors. The glucans produced by *Streptococcus mutans, Streptococcus sanguis* and *Leuconostoc mesenteroides* are made up from glucose units joined mainly by $\alpha 1{\rightarrow}6$ linkages with a variable number of branch points introduced by $\alpha 1{\rightarrow}3$ linkages. These polymers are synthesised directly from sucrose by an enzyme called glycosyl transferase or dextransucrase. The enzyme system, which can be bound to the cells or completely cell-free, transfers a glucose unit directly from sucrose to the 6-position of the glucose terminal of the growing dextran chain without the involvement of any intermediates. The glucans or dextrans are either released from the cells as 'water soluble' polymer or remain loosely attached in a gelatinous 'water insoluble' form as a capsule.

Structure of the Gram-positive Cell Wall

Having described the chemical structures of the main wall components and briefly discussed their biosynthesis we are in a position to speculate on how they are assembled and how they interact in the intact bacterial cell. Is it possible to relate the information which comes from the chemical analysis of cell walls to the structures seen in the electron microscope (Figure 2.1)? When viewed in thin transverse sections, Gram-positive walls do not normally exhibit any layering or fine structure. Although variation in the density of material in wall sections may sometimes be seen there is no evidence of segregation of the major components, peptidoglycan, tiechoic or teichuronic acids into discrete bands. It seems that these components form a mixed matrix which is depicted in Figure 2.11 as a cross-linked

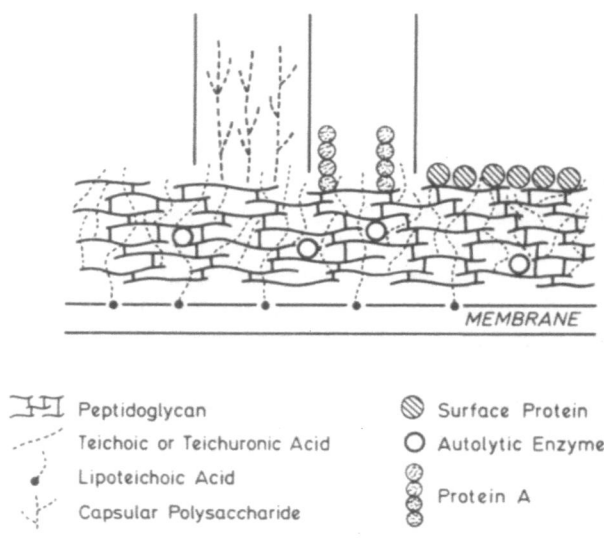

Peptidoglycan Surface Protein

Teichoic or Teichuronic Acid Autolytic Enzyme

Lipoteichoic Acid

Capsular Polysaccharide Protein A

Figure 2.11: General Structure of the Gram-positive Cell Wall Showing Possible Arrangement of Polymers

network of peptidoglycan interwoven with long, flexible chains of teichoic or teichuronic acids. The structure carries a large net negative charge, due chiefly to the phosphate groups in teichoic acid or the carboxyl groups in teichuronic acid. Peptidoglycan can also contribute a number of negative charges from any uncross-linked peptides, i.e. the free carboxyl groups of both the terminal D-alanine residues and the amino acid occupying the third position of the oligopeptide chain on muramic acid. The overall negative charge of the wall is partly reduced by the amino groups of alanyl substituents of teichoic acid chains. Each negatively charged group in the wall normally has a positively charged metal ion associated with it. In most growth media or natural environments these are most likely to be sodium and magnesium ions. Teichoic acids, in particular, have a strong, selective affinity for magnesium ions and one of their functions is thought to be to maintain an adequate supply of magnesium for the cell. One consequence of the charged, flexible nature of the teichoic and teichuronic acid polymers is their property of changing shape according to the ionic conditions. In low ionic strength media the negatively charged centres spaced regularly along the polymer chains tend to repel each other with the result that the polymers have extended rod-like conformations. In high ionic strength media the negative charges are effectively shielded by the high local concentration of metal ions and the polymers are able to adopt a more compact configuration. The implications of such shape changes can only be guessed at, but it seems reasonable to suppose that the wall is capable of altering its physical structure in response to changes in the ionic composition of the media.

The location of neutral polysaccharides and proteins within the walls is unknown; presumably they are embedded within the peptidoglycan-teichoic-teichuronic acid matrix as depicted in Figure 2.11. Protein A in *Staphylococcus aureus* is known to protrude from the wall into the surrounding medium. The polyglycerol phosphate chains of membrane teichoic acids are detectable on the cell surface and must therefore protrude through the wall from the cytoplasmic membrane. There is no evidence for a direct covalent linkage of capsular material to any other wall component. Association of capsules with the wall is usually sufficiently firm for the capsule to remain with the cells during removal from a liquid medium by centrifugation, although mechanical agitation such as subjecting them to gentle shearing in a hand homogeniser will remove some of the capsular material. In the case of *Streptococcus mutans* it seems that the cell-bound 'insoluble' form of the glucan polymer remains attached to the enzyme system (glycosyl transferase) which produces it on the cell surface.

Assembly of the Wall during Growth and Division

In streptococci, which grow as chains of spherical cells, new wall material is laid down at the septum, the site where the cells are eventually to divide. The new wall is fed out from the septum, pushing apart the two hemispheres of the original cell. At this stage the band of newly synthesised wall forms a distinctive V-shaped furrow around the cells. As the process continues the initially straight-sided new wall in the furrow begins to curve and two cells are formed. The plane of cell division is always the same so that regular chains are formed. In staphylococci the process is a little different. New cell wall is synthesised at the site of division but is not fed away from the site of deposition to form new peripheral walls. Instead the septum grows across the cell forming a thick cross wall, or equatorial band which divides the cell in two. The cross wall then peels apart and the two daughter cells round up. The next site of division is not oriented in the same plane, so the familiar grape-bunch clusters are formed after successive rounds of cell division.

The mechanism by which the wall is laid down during division of rod shaped organisms has proved to be difficult to study because the walls of Gram-positive rods undergo extensive turnover during growth. That is, some wall material (peptidoglycan and the accessory polymers), is degraded and released from the walls during growth, presumably as the

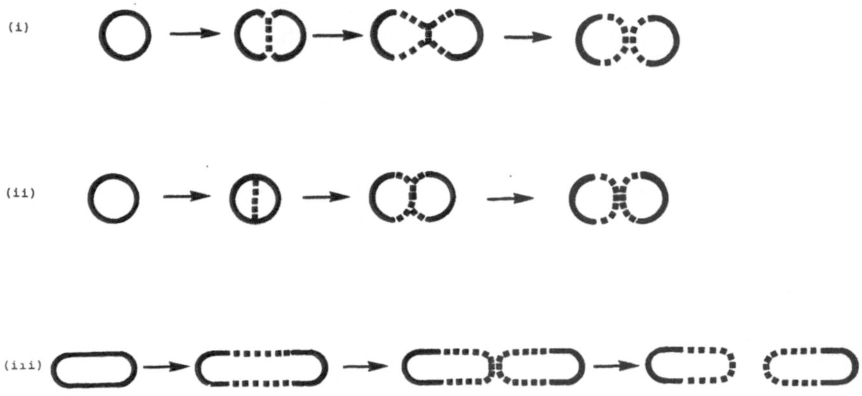

Figure 2.12: Wall Synthesis and Cell Division: (i) Streptococci, (ii) Staphylococci, (iii) Bacilli

result of the action of autolytic enzymes. Studies so far indicate that a rod-shaped organism lengthens by insertion of new wall at multiple sites along the side walls. The new wall material becomes integrated with the existing wall but, at the same time some wall is formed by ingrowth of the wall midway along the cell. Division takes place after the septum has grown across the cell to form two daughter cells. During the whole process of elongation and division wall material making up the end caps of the cells is conserved; no new wall seems to be deposited in this region and it is not subject to turnover. An outline of the processes involved in the division of streptococci, staphylococci and bacilli is shown in Figure 2.12.

We are still not in a position to say how, exactly, the newly synthesised wall polymers are inserted into the existing wall. The length of the glycan strands of peptidoglycan in bacilli (0.1–0.3 µm) is such that they must be oriented more or less horizontally along the length of the cell or around the circumference of the cylinder, rather than radially across the wall, which is only 0.03–0.08 µm thick. Addition of new glycan strands by transglycosylation to existing strands oriented horizontally would obviously extend the length of the cell. Since bacilli grow longer prior to division but do not increase their cylindrical diameter the extension must occur entirely along the length of the cells. However, the glycan strands might well be arranged in some intermediate orientation within the walls, in which case the formation of cross links in the peptidoglycan by transpeptidation would also contribute to the extension of the wall.

Study of the penicillin binding proteins has shown that bacteria contain several different enzymes (transpeptidases, endopeptidases and carboxy-peptidases) each involved with a separate aspect of peptidoglycan assembly. Although most of this work has been done with the Gram-negative organism *E. coli*, Gram-positives also have multiple penicillin-binding proteins. In both Gram-positive and Gram-negative bacteria the penicillin-binding proteins are enzymes which control how and where peptidoglycan is deposited in the wall and are intimately involved in controlling the overall shape of the organism. To maintain its characteristic morphology an organism must exert control over the various enzymes which influence its shape. The balance is normally controlled in a highly efficient manner so that cells continue to grow and divide over many generations, producing progeny of identical shape. Occasionally the control is lost and aberrant morphologies result: some mutants of the normally rod-shaped *B. subtilis* grow as spheres, some as helically wound filaments and others as tight corkscrew-like helices.

Further Reading

Amako, A. and Umeda, A. 'Bacterial Surfaces as Revealed by the High Resolution Scanning Electron Microscope', *Journal of General Microbiology* (1977), *98*, 297–9

Anderson, A.J., Green, R.S., Sturman, A.J. and Archibald, A.R. 'Cell Wall Assembly in *Bacillus subtilis*, Location of Wall Material Incorporated During Pulsed Release of Phosphate Limitation, its Accessibility to Bacteriophages and Concanavalin A, and its Susceptibility to Turnover', *Journal of Bacteriology* (1978), *136*, 886–99

Archibald, A.R. 'The Structure, Biosynthesis, and Function of Teichoic Acid', *Advances in Microbial Physiology* (1974), *11*, 53–95

Archibald, A.R. and Coapes, H.E. 'Bacteriophage SP50 as a Marker for Cell Wall Growth in *Bacillus subtilis*', *Journal of Bacteriology* (1976), *125*, 1195–206

Berkeley, R.C.W., Lynch, J.M., Melling, J., Rutter, P.R. and Vincent, B. (eds). *Microbial Adhesion to Surfaces* (Ellis Horwood Ltd, Chichester, 1980)

Birdsell, D.C., Doyle, R.J. and Morgenstern, M. 'Organization of Teichoic Acid in the Cell Wall of *Bacillus subtilis*, *Journal of Bacteriology* (1975), *121*, 726–34

Burdett, I.D.J. and Higgins, M.L. 'Study of Pole Assembly in *Bacillus subtilis* by Computer Reconstruction of Septal Growth Zones Seen in Central Longitudinal Thin Sections of Cells', *Journal of Bacteriology* (1978) *133*, 959–71

Coley, J., Tarelli, E., Archibald, A.R. and Baddiley, J. 'The Linkage Between Teichoic Acid and Peptidoglycan in Bacterial Cell Walls', *FEBS Letters* (1978), *88*, 1–9

Doyle, R.J., McDannel, M.L., Helman, J.R. and Streips, U.N.

'Distribution of Teichoic Acid in the Cell Wall of *Bacillus subtilis*', *Journal of Bacteriology* (1975), *122*, 152–8

Hamada. S. and Slade, H.D. 'Biology, Immunology and Cariogenicity of *Streptococcus mutans*', *Microbiological Reviews* (1980), *44*, 331–84

Hussey, H., Sueda, S., Cheah, S.C. and Baddiley, J. 'Control of Teichoic Acid Synthesis in *Bacillus licheniformis* ATCC 9945', *European Journal of Biochemistry* (1978), *82*, 169–74

Pooley, H.M. 'Turnover and Spreading of Old Wall During Surface Growth of *Bacillus subtilis*', *Journal of Bacteriology* (1976), *125*, 1127–38

Rogers, H.J., Perkins, H.R. and Ward, J.B. *Microbial Cell Walls and Membranes* (Chapman and Hall, London, 1980)

Sargent, M.G. 'Surface Extension and the Cell Cycle in Prokaryotes', *Advances in Microbial Physiology* (1978), *18*, 106–76

Wyke, A.W. and Ward, J.B. 'Biosynthesis of Wall Polymers in *Bacillus subtilis*', *Journal of Bacteriology* (1977), *130*, 1055–63

3 THE ENVELOPE OF GRAM-NEGATIVE BACTERIA

Examination of high-resolution electron micrographs of sections through Gram-negative and Gram-positive bacteria reveals gross morphological differences. The cytoplasmic membrane of Gram-positive bacteria is encased with a thick peptidoglycan layer and in some species an extra-cellular capsule (Figure 3.1). The envelope of Gram-negative bacteria is more complex. The cell wall peptidoglycan appears to be less substantial and not as closely associated with the cytoplasmic membrane as the equivalent structure in Gram-positive bacteria. The most notable feature of the Gram-negative envelope is the presence of a second membrane to the exterior of the peptidoglycan: the outer membrane (Figure 3.1). Both membranes consist of lipid/protein bilayers, of the typical Davson-Danielli model. However, beyond this superficial resemblance, few structural or physiological similarities between the cytoplasmic and outer membranes exist, reflecting the differing roles of the membranes in the life of the bacterium.

The cytoplasmic membrane contains the enzyme systems of the electron transport chain (cytochromes) and oxidative phosphorylation, systems for the active transport of solutes and excretion of waste products plus the synthetic apparatus necessary for the production of exterior layers. The outer membrane of Gram-negative bacteria has no role in electron

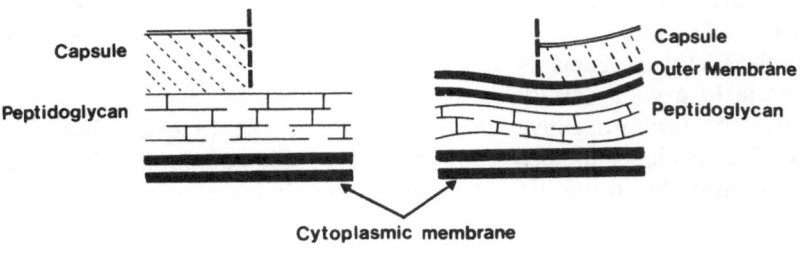

GRAM POSITIVE   **GRAM NEGATIVE**

Figure 3.1: Diagrammatic Representation of the Cell Surface of Gram-positive and Gram-negative Bacteria

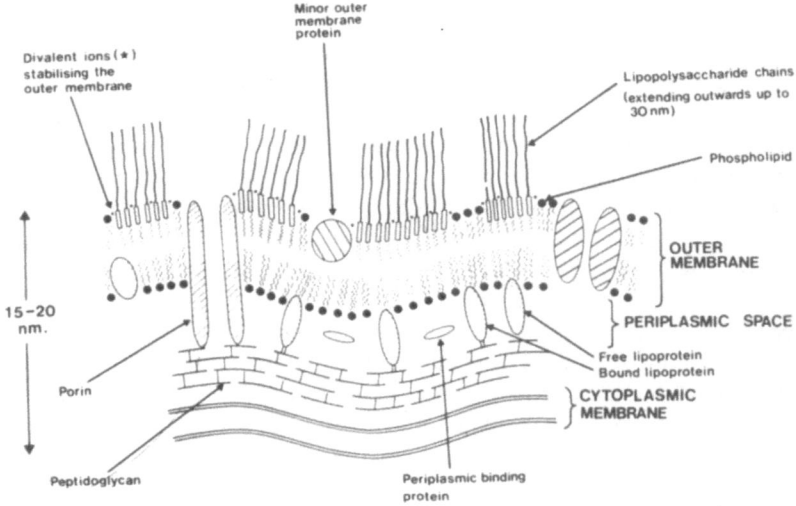

Figure 3.2: Representation of the Envelope of Gram-negative Bacteria (cytoplasmic membrane not drawn to scale)

transport and extremely limited enzymic activity. It is distinguished by the presence of a novel series of proteins and lipopolysaccharide (LPS); molecules unique to the envelope of Gram-negative bacteria (Figure 3.2). This outer membrane confers special properties upon Gram-negative bacteria. The outer membrane constitutes a barrier making the surface of Gram-negative organisms less permeable than that of Gram-positive bacteria to a wide variety of molecules. In general the outer membrane makes the Gram-negative envelope impermeable to hydrophobic compounds and higher molecular weight hydrophilic compounds. The restrictive permeability properties of the outer membrane account for the resistance of Gram-negative bacteria to a wide range of antibiotics and antimetabolites, preventing the accumulation of lethal concentrations of the agent at its active site. The outer membrane is also important to the growth of the organism in an animal host; allowing the bacterium to survive the emulsifying action of bile salts and enzymes of the gut, and aids resistance to host defence mechanisms, for example phagocytosis by white blood cells and the complement-directed bactericidal action of serum (see Chapter 5).

The barrier function of the outer membrane first became amenable to study, when it was found that treatment with ethylenediaminetetra-acetic acid (EDTA) markedly increased the permeability of the envelope. Treatment of *Escherichia coli* with low levels of EDTA can produce cells which are viable, retain normal growth rates in which protein, RNA and DNA syntheses remain unaffected; phage absorption is unchanged and cytoplasmic membrane funtions (i.e. active transport, energy transduction and biosyntheses) continue unabated. However, EDTA-treated bacteria become vulnerable to a wide range of antimicrobial agents to which they were previously insensitive. Additionally the membranes and walls of EDTA-treated bacteria are respectively sensitive to phospholipase and lysozyme. Enzyme substrates normally excluded from Gram-negative cells, e.g. adenosine triphosphate, carbamoyl phosphate and ribonucleotide triphosphates are taken up by EDTA-treated bacteria. When Gram-negative bacteria are subjected to osmotic shock in the presence of EDTA a series of soluble proteins are released. The proteins are normally confined between the two membranes (the periplasmic space) and are hence known as periplasmic proteins.

These observations are consistent with the hypothesis that EDTA causes a non-specific increase in the general permeability of the outer membrane. EDTA is a strong chelator, i.e. has a high affinity for simple ions, particularly divalent ions such as magnesium and calcium. Indeed the effect of EDTA upon the envelope may be reversed by addition of excess divalent ions. EDTA-treatment of enteric bacteria, e.g. *E. coli* and *Salmonella* species, releases between a third and a half of the LPS component of the outer membrane, together with some (5–20 per cent) of the phosphatidylethanolamine and a little protein. Since the addition of divalent ions prevents the loss of this material it is thought that divalent ions, particularly Mg^{2+}, are important in stabilising protein-LPS and phospholipid-LPS interactions in the outer membrane. *Pseudomonas aeruginosa*, and other pseudomonads, differ from the enteric bacteria in that they are supersensitive to EDTA; extremely low concentrations causing lysis of the bacterium. This suggests that the outer membrane of pseudomonads may differ significantly from that of other Gram-negative bacteria.

In spite of the great structural and physiological diversity found amongst Gram-negative bacteria they appear to share a common pattern in the structure of their outer membranes. Determination of the composition of the outer membrane must be preceded by separation from other cellular components, in particular from contamination by cell wall and cytoplasmic membrane fragments. In general the bacteria are broken open using a pressure cell, treated with lysozyme to dissolve away the peptidoglycan and then washed free of cytoplasmic debris. The remaining lipid-rich material consisting of both cytoplasmic and outer membranes, must then be separated either by treatment with the detergent *N*-lauroyl sarcosine, which selectively solubilises the cytoplasmic membrane or by using sucrose density ultracentrifugation which resolves the membranes into two distinct fractions on the basis of their differing densities.

The membrane fractions are then washed. The efficiency of the proce-

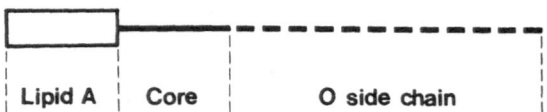

Figure 3.3: Representation of the Three Regions of Bacterial Lipopolysaccharides (LPS). The lipid A region is embedded in the phospholipid bilayer, the core and O-side chain polysaccharide extending outwards from the cell

dure in isolating pure outer membrane can be gauged by examining for the presence of cytoplasmic membrane contamination. Pure outer membrane should be demonstrably free of the activity associated with enzymes known to be firmly bound to the cytoplasmic membrane, e.g. lactate dehydrogenase, succinic dehydrogenase and cytochromes. Analysis of isolated outer membrane shows it to consist of three major components; LPS, protein and phospholipid. Estimates of the number of molecules of these components present in 1 μm^2 section of outer membrane suggest that there are 10^5 molecules of LPS, 10^5 molecules of protein and 10^6 molecules of phospholipid. This diversity of molecular types confers upon the outer membrane a series of unique structural and functional features.

Lipopolysaccharide (LPS)

The term lipopolysaccharide denotes any molecular species that consists in the main of polysaccharide, but also possess a significant lipid moiety. In bacteriology the term has acquired an especial connotation, being exclusively applied to the endotoxic O-antigens from 'smooth' Gram-negative bacteria and the corresponding products from 'rough' strains or mutants. The terms 'rough' and 'smooth' refer to the colonial morphology of the strain on nutrient-rich solid media, the former appearing crenated with an irregular boundary, the latter possessing an unwrinkled and more rounded glossy appearance. The classic model for bacterial LPS is that originally proposed by Luderitz and Westphal for the material extracted by aqueous phenol from smooth (wild-type) strains of *Salmonella*. Figure 3.3 represents the structural unit of LPS, and shows it to be composed of three covalently linked regions, each with a distinctive composition and biological function. Integrated into the outer membrane bilayer is the lipid A region, consisting of fatty acid chains linked to glucosamine. Extending outwards from the lipid A is the core region consisting of a short carbohydrate chain and from the core, in smooth strains, a longer carbohydrate polymer, the O-side chain.

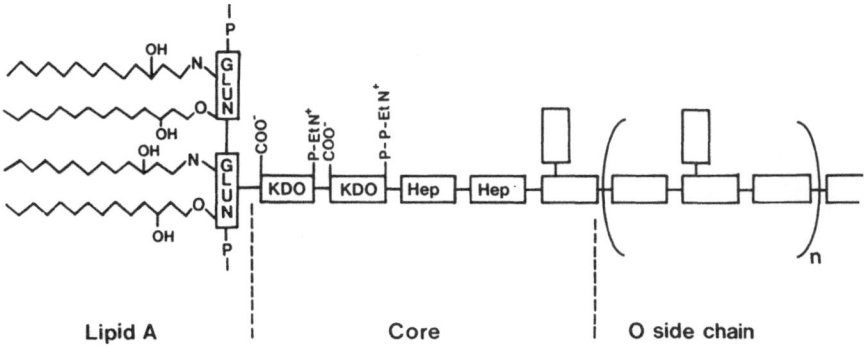

Figure 3.4: Generalised Structure of Bacterial Lipopolysaccharide. The structure of the lipid A region is highly conserved amongst a wide range of Gram-negative bacteria. Similarly little variation is found in the inner core (KDO-heptose) region. Considerable variation is present in the monosaccharides that constitute the outer core and O-side chain regions and consequently they have been left as empty blocks.

LPS may be extracted from intact cells by a variety of procedures. Treatment with chelating agents, such as EDTA, releases about one-third of the LPS. Generally LPS is obtained by a phenol-water extraction. In essence the bacterial cells are treated for 1h with 45 per cent (w/v) aqueous phenol at 68°C. On cooling the mixture separates into two phases, a lower phenol layer and an upper aqueous phase, in which the LPS is located. This treatment extracts up to 90 per cent of the LPS which can then be subjected to chemical analysis.

Structure of LPS

Figure 3.4 shows a structure typical of LPS isolated from *Salmonella* species. The composition of lipid A is relatively invariable, its structure being highly conserved in a wide range of Gram-negative bacteria. The sugars comprising the R core are similar in many different bacterial species. In contrast the polysaccharide of the O-side chains where present, exhibit gross differences in composition and structure, often within a single species.

Lipid A. The ketosidic linkage between the terminal core sugar 3-deoxy-D-*manno*-octulosonic acid (KDO) and lipid A is acid labile. Hence the hydrophobic lipid A region can be easily separated from isolated LPS by mild acid hydrolysis, followed by chloroform extraction. The structural basis of the LPS extracted from enteric bacteria is a glucosamine disacch-

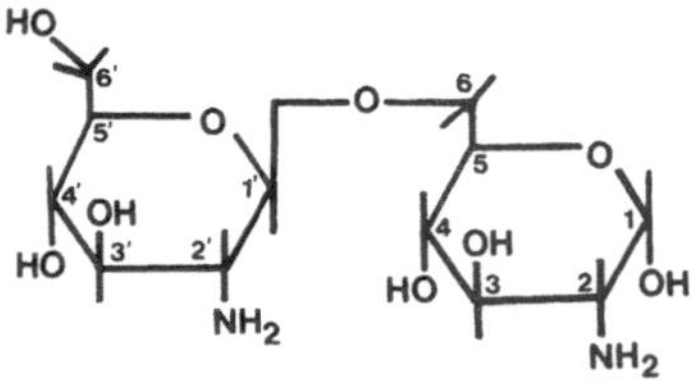

Figure 3.5: Structure of the Diglucosamine Backbone of Lipid A from *Escherichia coli*

aride; in *Salmonella, Shigella, Serratia* and *Pseudomonas* species this is a
$\beta1\rightarrow6$ linked glucosamine 1-D-glucosamine disaccharide (Figure 3.5). In
some other species, for example *E. coli* 086 and *Shigella flexneri*, the
linkage between glucosamine residues is $\beta1\rightarrow4$ not $\beta1\rightarrow6$. In salmonellae the
glucosamine disaccharide contains phosphate groups at C-1 and C-4' and
is linked through its 6'-hydroxyl group ketosidic bond to the terminal KDO
of the R core (Figure 3.6). Lipid A is unique among bacterial lipids in
containing hydroxy fatty acids, almost invariably 3-D-hydroxy fatty acids.
In addition the lipid A contains medium-length straight chain fatty acids.
Much controversy exists on the exact number and positioning of the fatty
acids in the lipid A of salmonellae and *E. coli*. However the picture
emerging is that only the 3-OH fatty acids, usually β-hydroxy-myristic acid,
directly substitute the glucosamine disaccharide either amide linked to C-2'
and C-2 or ester-linked to C-3'. The myristic acid may be 3-*O*-acetylated
by either a straight chain or another 3-OH fatty acid (Figure 3.6). Similar
lipid A structures are present in many other Gram-negative bacteria
including *Shigella, Pseudomonas, Fusobacteria* and *Selenomonas* species.

The lipid A of many purple non-sulphur bacteria does not contain
glucosamine, but its place is taken by a diaminohexose, 2,3-diamino-2,3-
dideoxy-D-glucose in species such as *Rhodopseudomonas palustris* and
R. viridis (Figure 3.7). In *R. viridis*, the amide groups of the amino sugars
are acylated with β-hydroxymyristic acid; the hydroxyl groups remaining
unsubstituted. Lipid A from *Rhodospirillum tenue* contains a unique
carbohydrate configuration, consisting of two amino sugars: D-glucosamine, 4-amino-L-arabinose and the neutral sugar D-arabinose, to which is
linked a third non-*N*-substituted D-glucosamine attached to C-4 of the
central disaccharide (Figure 3.8a). The phosphate at the reducing end of
the central glucosamine disaccharide is substituted by a neutral sugar
D-arabino furanose. The external sugar residues (i.e. 4 amino-L-arabinose,
D-arabinose and the third D-glucosamine) are not acylated (Figure 3.8b).

A variety of straight chain and D-3-OH fatty acids may be present in the

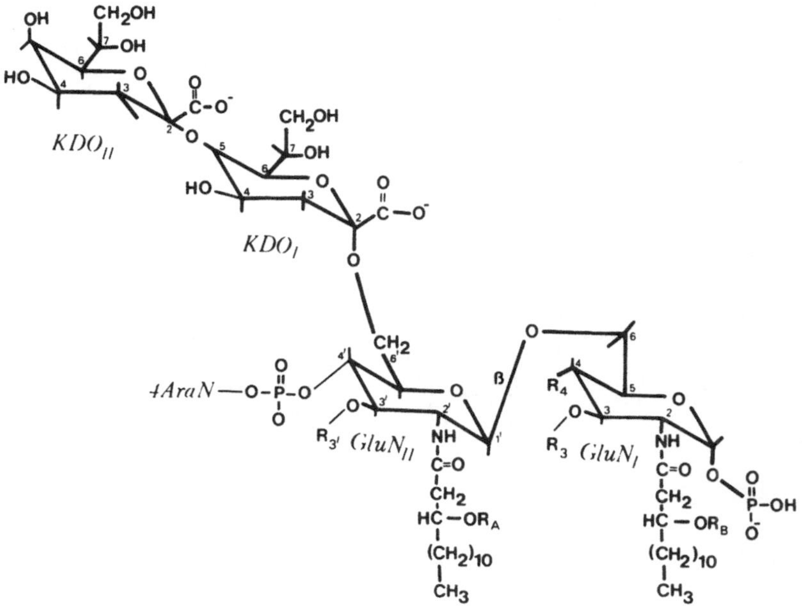

Figure 3.6: Structure of the Lipid A of *Salmonella* Species. The two glucosamine sugars are amide-linked at the 2-position to 3-OH fatty acids, which may themselves bear ester-linked fatty acids (R_A, R_B). Additional acylation by ester-linked 3-OH fatty acids may occur at 3 and 3' positions. The sugar 4-amino-arabinose may be substituted at the 4' of the glucosamine II and ethanolamine at the 1 position of glucosamine I. The KDO disaccharide (not strictly part of lipid A) is attached to glucosamine II 6'. The 4' position of glucosamine I is probably unsubstituted

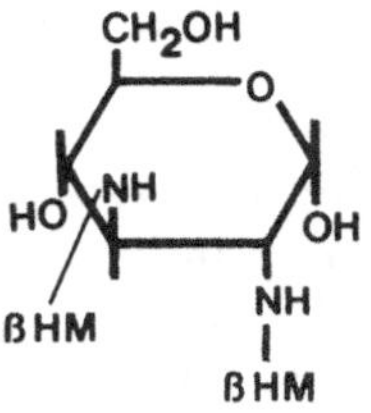

Figure 3.7: Structure of the Lipid A of *Rhodopseudomonas viridis*. In this organism a 2,3-dideoxy-D-glucose monomer replaces the glucosamine dimer as the basis of the lipopolysaccharide. The molecule bears two amide-linked 3-OH fatty acid substituents

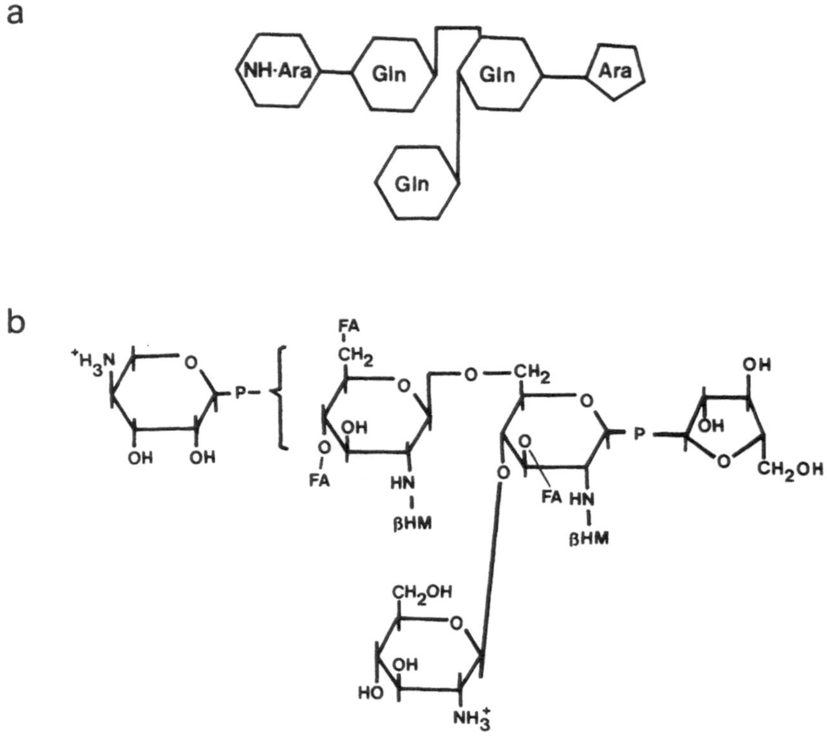

a

b

Figure 3.8: Structure of Lipid A from *Rhodospirillum tenue*. (a) The basis of the sugar backbone; (b) the complete lipid A molecule. (Gln; glucosamine, Ara; arabinose, NH.Ara; 4-amino-arabinose, βHM; β-hydroxymyristic acid and FA; 14:0 or 16:0 or D-3-OH 10:0 fatty acid.)

lipid A of Gram-negative bacteria (Table 3.1). In general the amide- and ester-linked 3-OH fatty acids are straight chained, even numbered and saturated, although some odd numbered and iso-branched hydroxy fatty acids have been detected.

Amongst the enterobacteria a single type of hydroxy fatty acid is linked to the glucosamine disaccharide and this is usually β-hydroxymyristate. However in other species, for example *Xanthomonas*, *Veillonella* and myxobacteria, two different 3-OH fatty acid species are amide-linked to glucosamine. The 3-OH fatty acid represents between one-half and three-quarters of the total fatty acids in enterobacterial LPS. However in *Klebsiella aerogenes*, hydroxy fatty acids only constitute 6 per cent of the total, while in *Brucella* none can be detected. Lipid A of *Pseudomonas* contains additional 2-hydroxy fatty acids. The non-hydroxy fatty acids present in enterobacterial LPS are also saturated, straight chain and even numbered, including lauric (C-12), myristic (C-14) and palmitic (C-16)

Table 3.1: Major Fatty Acid Composition of Bacterial LPS. The hydroxy fatty acids may be amide- or ester-linked to glucosamine. The non-hydroxy fatty acids may be only ester-linked to the 3-OH fatty acids.

Isolated LPS	D-3-hydroxy fatty acids	Non-hydroxy fatty acids
Rhodospirillum tenue	3-OH C10:0	?
Neisseria gonorrhoeae	3-OH.C10:0	C12:0
	3-OH C12:0	C14:0
	3-OH C14:0	C18:0
Neisseria meningitidis	3-OH C12:0	C12:0
	3-OH C14:0	
Salmonella species ⎫		⎧ C12:0
Escherichia coli ⎬	3-OH C14:0	⎨ C14:0
Serratia marcescens and *Yersinia* species ⎭		⎩ C16:0
Rhodopseudomonas viridis	3-OH C14:0	None present
Rhodopseudomonas palustris	3-OH C14:0	C16:0
		C18:0
		C18:1
Veillonella species	3-OH C13:0	C13:0
	3-OH C15:0	
Myxobacteria	3-OH *i*-C15:0	*i*-C15:0
	3-OH *i*-C17:0	
Xanthomonas	3-OH C12:0	C10:0
	3-OH *i*-C13:0	*i*-C11:0
Pseudomonas aeruginosa	2-OH C12:0	C12:0
	3-OH C10:0	C16:0
	3-OH C12:0	

acids. In *E. coli, Proteus, Salmonella* and *Shigella* strains, the hydroxy fatty acids are 3-O-acylated to produce 3-myristoxy myristic fatty acid (Figure 3.6). In *Vibrio parahaemolyticus* a similar fatty acid 3-lauroxy lauric acid is present. In other species, for example *Xanthomonas, Rhodopseudomonas capsulata* and myxobacteria, the 3-hydroxy acids remain unsubstituted. The 2,3-diamino-2,3-dideoxy-D-glucose of lipid A from *R. viridis* is unusual in that it does not seem to possess any ester-linked fatty acids. The predominance of straight chain saturated fatty acids in bacterial LPS encourages close packing of the acyl chains. This is thought to be important in the integration of the LPS into the phospholipid bilayer.

Determination of the molecular weight of acetylated LPS by ultracentrifugation in acetone or by chromatography, gives a molecular weight approximately three times greater than expected. This has been explained in terms of a cross linking of, on average, three monomeric units within the lipid A region. Some early evidence indicated that in *E. coli* and *S. minnesota* lipid A monomers were linked through phosphate bridges. However more recent studies, particularly those using ^{31}P nuclear magnetic resonance spectroscopy, failed to detect the presence of pyrophosphate bridges. More studies are needed but it appears that the LPS oligomers are not covalently linked by phosphate bridges but may be held together by divalent ions or LPS-associated proteins. About a quarter of the 4'-phosphate groups of the lipid A of *Salmonella* are linked to the 1-position of amino-arabinose and an approximately equal proportion of the reducing terminal of the lipid A to ethanolamine (Figure 3.5). The lipid A of *Rhodospirillum* also contains similar substituents (Figure 3.8) but their presence cannot be detected in deep rough mutants of *E. coli* K12.

Core. The core is a complex oligosaccharide. In the vast majority of Gram-negative bacteria the core sugars are linked to lipid A by the sugar acid 3-deoxy-D-*manno*-2-octulosonate (KDO). It was first thought that the core of enteric bacteria contained a KDO trimer, but it now appears that this was an overestimate, and that the inner core region contains a $\alpha 2 \rightarrow 6$ linked KDO dimer (Figure 3.6). The KDO region of the LPS appears to be indispensible, since mutants defective in the biosynthesis of the inner core, i.e. $(KDO)_2$-region, have never been isolated, although mutants with defects in the outer core, that is deep rough mutants, are common. The core of *S. typhimurium* has been studied in great detail; the determination of structure and biosynthesis being greatly facilitated by the use of mutants defective in core biosynthesis. The core of *S. typhimurium* consists of *N*-acetyl-D-glucosamine, D-glucose, D-galactose and L-glycero-D-mannoheptose (Figure 3.9). In addition the core contains phosphate groups, *O*-phosphorylethanolamine and *O*-pyrophosphorylethanolamine. The core regions of most *Salmonella* species are very similar but in *E. coli* more than one core type has been identified. The structure of the core of *E. coli* K12 is very similar to that of *Salmonella* (Figure 3.10). The R1 and R2 core types of *E. coli* (Figures 3.11 and 3.12 respectively) could be considered as variations of the general theme. The core of *E. coli* B contains only glucose, heptose and KDO (Figure 3.13) leading to the suggestion that this strain lacks some of the genes necessary for the synthesis of a complete R core. Analysis of the core structure of *Shigella* species (Figure 3.14) shows them to be very similar to *E. coli* and that like *E. coli*, many *Shigella* species share related core configurations. The core of *Pseudomonas aeruginosa* differs significantly from that of other species, containing KDO, glucose, heptose, rhamnose, galactosamine and alanine.

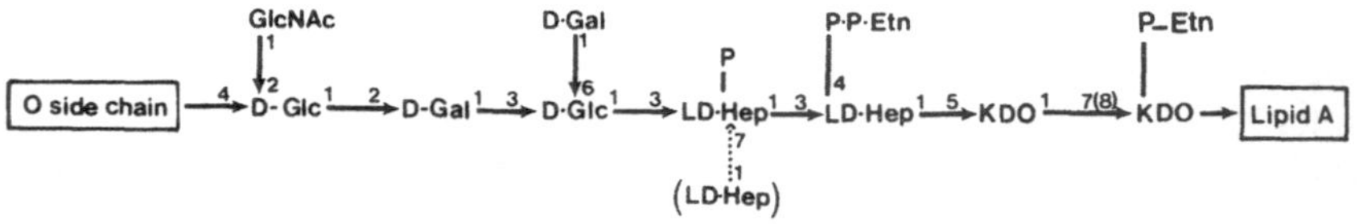

Figure 3.9: Core Region of *Salmonella typhimurium* Lipopolysaccharide. (Glc, glucose; Gal, galactose; Hep, heptose; GlcNAc, acetylglucosamine; KDO, 3-deoxy-D-manno-octulosonic acid; and Etn, ethanolamine.)

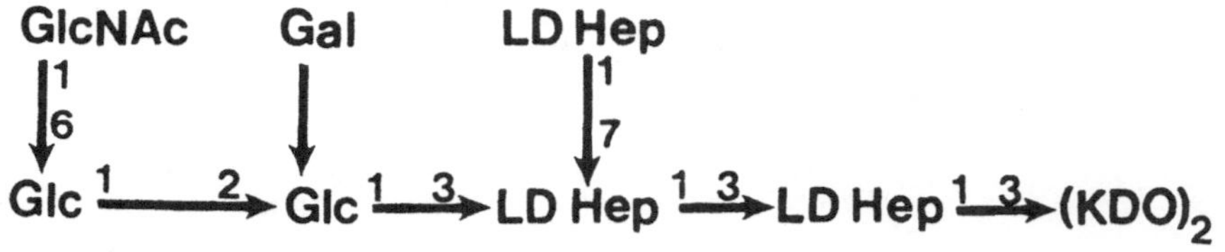

$$\text{GlcNAc} \qquad \text{Gal} \qquad \text{LD Hep}$$

$$\text{Glc} \xrightarrow{\;1\qquad2\;} \text{Glc} \xrightarrow{1\quad3} \text{LD Hep} \xrightarrow{1\quad3} \text{LD Hep} \xrightarrow{1\quad3} (\text{KDO})_2$$

Figure 3.10: Core of *E. coli* K12 LPS

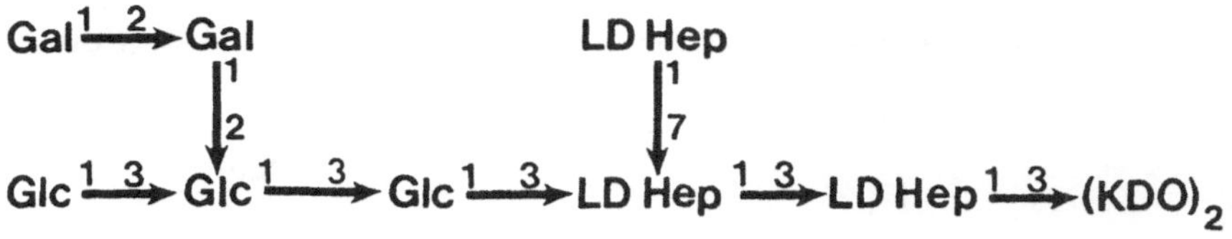

$$\text{Gal} \xrightarrow{1\quad2} \text{Gal} \qquad\qquad \text{LD Hep}$$

$$\text{Glc} \xrightarrow{1\quad3} \text{Glc} \xrightarrow{1\quad3} \text{Glc} \xrightarrow{1\quad3} \text{LD Hep} \xrightarrow{1\quad3} \text{LD Hep} \xrightarrow{1\quad3} (\text{KDO})_2$$

Figure 3.11: R1-core Structure of *E. coli* LPS

$$\text{GlcNAc} \qquad\qquad \text{Gal} \qquad \text{LD Hep} \qquad\qquad \text{Gal}$$

$$\text{Glc} \xrightarrow{1\quad2} \text{Glc} \xrightarrow{1\quad3} \text{Glc} \xrightarrow{1\quad3} \text{LD Hep} \xrightarrow{1\quad3} \text{LD Hep} \xrightarrow{1\quad3} (\text{KDO})_2$$

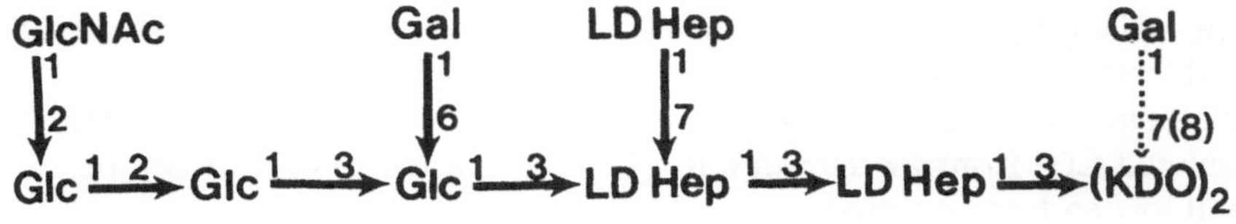

Figure 3.12: R2-core Structure of *E. coli* LPS

$$\overset{\text{P}}{\text{Glc} \xrightarrow{1\quad3} \text{Glc} \xrightarrow{1\quad3} \text{LD Hep} \xrightarrow{1\quad3} \text{LD Hep} \xrightarrow{1\quad5} (\text{KDO})_2}$$

$$\text{Hep.P-P- EtN}$$

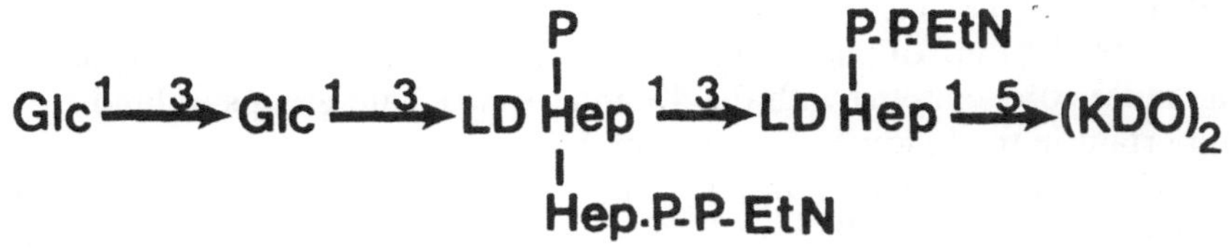

Figure 3.13: Core Structure of *E. coli* B LPS

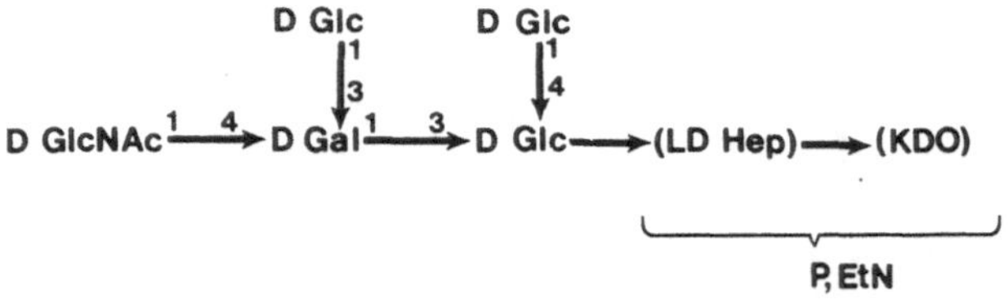

$$\begin{array}{ccc}
\text{D Glc} & \text{D Glc} \\
\downarrow^1_3 & \downarrow^1_4 \\
\text{D GlcNAc} \xrightarrow[]{1\ \ 4} \text{D Gal} \xrightarrow[]{1\ \ 3} \text{D Glc} \longrightarrow \text{(LD Hep)} \longrightarrow \text{(KDO)}
\end{array}$$

$$\underbrace{\qquad\qquad\qquad\qquad}_{\text{P, EtN}}$$

Figure 3.14: Core Structure of *Shigella* LPS

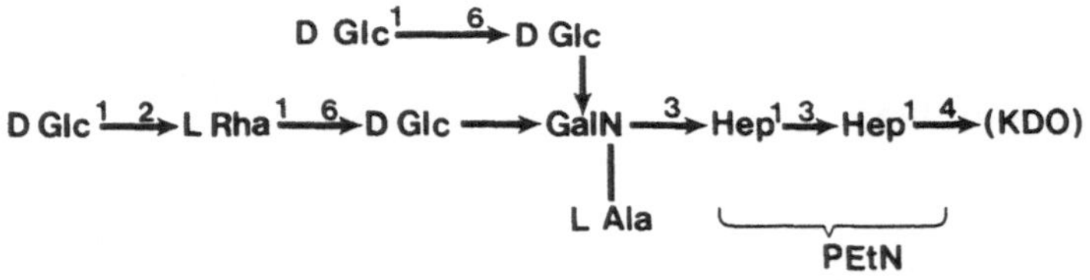

$$\begin{array}{c}
\text{D Glc} \xrightarrow[]{1\ \ 6} \text{D Glc} \\
\downarrow \\
\text{D Glc} \xrightarrow[]{1\ \ 2} \text{L Rha} \xrightarrow[]{1\ \ 6} \text{D Glc} \longrightarrow \text{GalN} \xrightarrow[]{3} \text{Hep}^1 \xrightarrow[]{3} \text{Hep}^1 \xrightarrow[]{4} \text{(KDO)} \\
\downarrow \\
\text{L Ala} \quad \underbrace{\qquad\qquad}_{\text{PEtN}}
\end{array}$$

Figure 3.15: Core Structure of *Pseudomonas aeruginosa* LPS. (Rha, rhamnose; GalN, galactosamine; Ala, alanine.)

In this species the heptose component consists of equal amounts of L-gly-cero-D-*manno*-heptose and D-glycero-D-*manno*-heptose (Figure 3.15). Analysis of the core of anaerobic Gram-negative bacteria, e.g. *Bacteroides* and certain photosynthetic bacteria, failed to detect KDO or heptose, suggesting that in these species another sugar links the core to lipid A. Isolated LPS from *Bacteroides* also has an unusual fatty acid composition and low endotoxic activity.

Although the outer part of the core shows considerable variation the heptose-KDO region remains highly conserved in most Gram-negative bacteria. The phosphorylethanolamine residues that substitute onto the inner core region contain free amino groups conferring zwitterionic character to the molecule. The adjacent ethanolamine and 4-amino-arab-inose substituents on the lipid A moiety are also zwitterions. These groups, taken as a whole may be important in creating a structure analogous to the polar 'head group' of phospholipids. The high proportion of ionic groups in the inner core, coupled with the 4'- and 1-phosphate groups of lipid A are important in the binding of cations by LPS.

O-side Chain Polysaccharides. The O-side chain is the serologically dominant part of the LPS molecule. It consists of repeating oligosaccharide units, often containing rare sugars and it is in these sugars that the sero-logical determinants reside. Polysaccharides are capable of a considerable degree of diversity per unit structure as the result of variation in the sugar composition and in the configuration of the glycosidic linkage. This

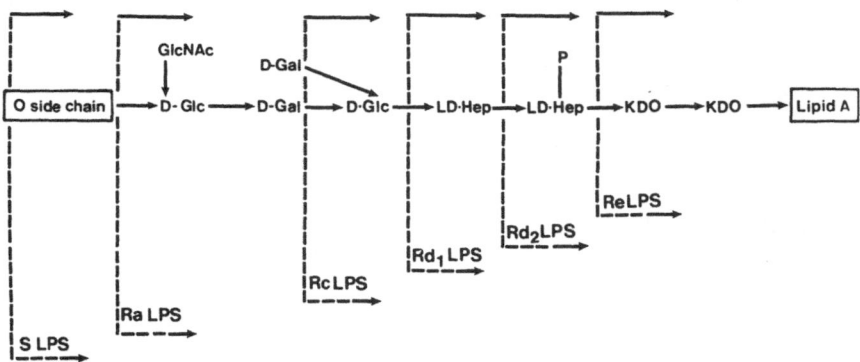

Figure 3.16: Structure of the Core Polysaccharides Produced by Rough Mutants (Ra to Re) of *Salmonella typhimurium*

accounts for the large number of O-serotypes in Gram-negative bacteria; up to one hundred have been described in *Salmonella* and an equivalent number in *E. coli.* Determination of the structure of the O-side chain polysaccharide has been achieved by the use of a variety of techniques, including immunochemical, biochemical, bacteriological and genetical studies. The structure is first detected and classified on the basis of its antigenicity. The O-side chain can be released from extracted LPS by mild acid hydrolysis and separated from other water-soluble hydrolysis products by gel-filtration techniques. The use of rough strains and mutants defective in different stages of LPS biosynthesis has proved invaluable in the determination of O-side chain structure. Smooth strains of enteric bacteria produce an LPS with a complete core and O-side chain; semi-rough mutants (*rfc* locus) appear unable to polymerise the O-specific oligosaccharide monomers, and rough mutants (*rfb*) cannot synthesise or attach the side chains. Rough mutants which produce defective cores (*rfa*) lack the ability to transfer the O-specific side chain from the carrier lipid (Figure 3.16). In some mutants the genetic defect is leaky and the strain can produce LPS bearing a few side chains (part-rough). The O-side chain is absent from mutants unable to synthesise component sugars. Galactose-epimerase-less mutants of *S. typhimurium* appear rough on minimal solid medium, since they cannot synthesise their galactose-containing O-side chain. If, however, they are supplied with exogenous galactose they appear smooth and can be shown to possess a complete LPS O-side chain. Although the viability of Gram-negative bacteria in laboratory culture seems unaffected by the loss of the O-side chain and the outer core region, these structures are of considerable importance for the ability of the organism to compete in natural habitats (see Chapters 5 and 6).

Bacterial O-side chains have a basic structural similarity, consisting of

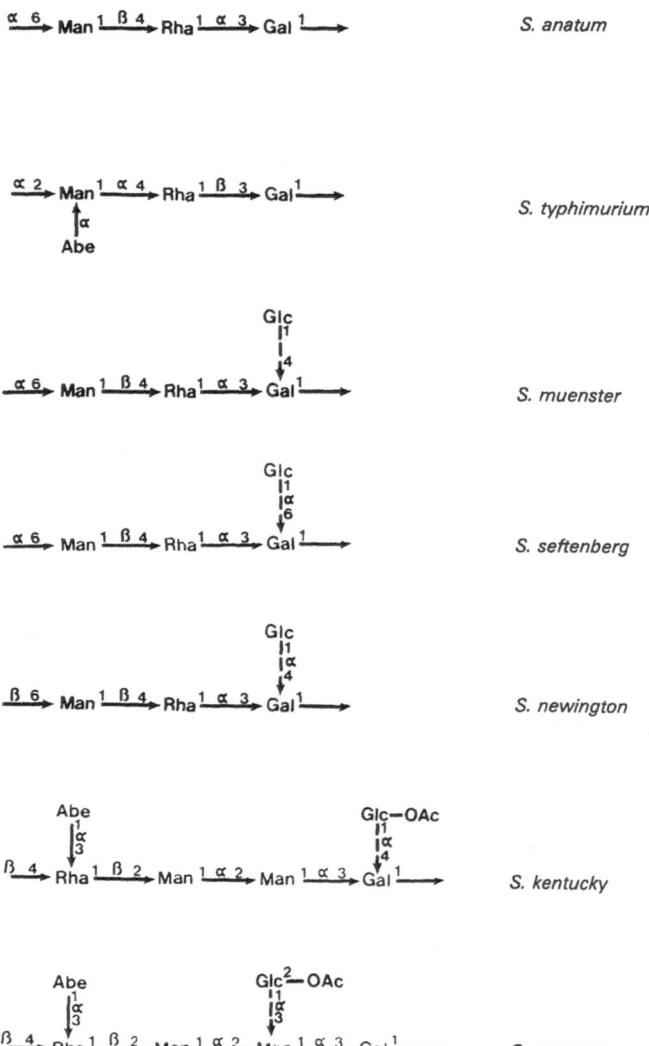

Figure 3.17: Structure of the Repeating Units of the O-antigen Oligosaccharide of LPS from *Salmonella* Species (Man, mannose; Abe, abequose.)

repeating sequences of usually two to four monosaccharides. The sugar constituents of the O-side chains can be determined using electrophoresis and chromatography, usually after hydrolysis of the polysaccharide. Analysis of the composition of large numbers of test strains permits the

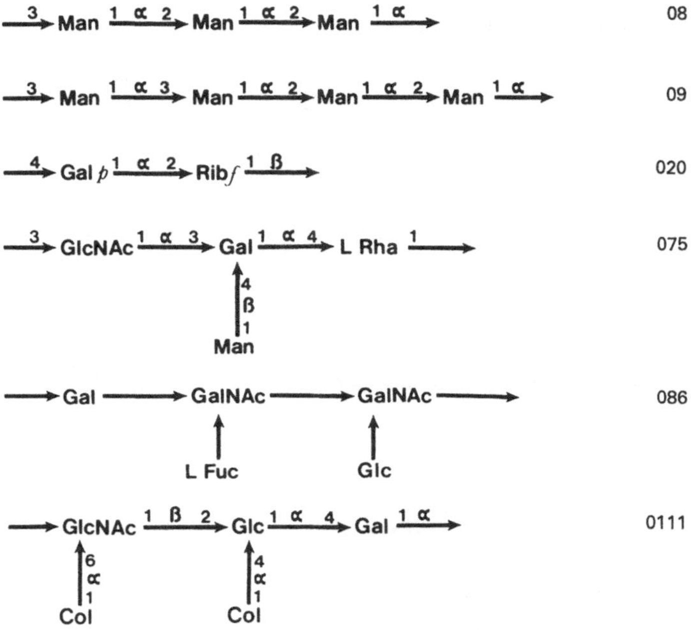

$$\xrightarrow{\ 3\ } \text{Man} \xrightarrow{1\ \alpha\ 2} \text{Man} \xrightarrow{1\ \alpha\ 2} \text{Man} \xrightarrow{1\ \alpha} \qquad\qquad 08$$

$$\xrightarrow{\ 3\ } \text{Man} \xrightarrow{1\ \alpha\ 3} \text{Man} \xrightarrow{1\ \alpha\ 2} \text{Man} \xrightarrow{1\ \alpha\ 2} \text{Man} \xrightarrow{1\ \alpha} \qquad 09$$

$$\xrightarrow{\ 4\ } \text{Gal}\, p \xrightarrow{1\ \alpha\ 2} \text{Rib}f \xrightarrow{1\ \beta} \qquad\qquad 020$$

$$\xrightarrow{\ 3\ } \text{GlcNAc} \xrightarrow{1\ \alpha\ 3} \text{Gal} \xrightarrow{1\ \alpha\ 4} \text{L Rha} \xrightarrow{1} \qquad 075$$

with branch:
$$\begin{array}{c} 4 \\ \beta \\ 1 \\ \text{Man} \end{array}$$

$$\longrightarrow \text{Gal} \longrightarrow \text{GalNAc} \longrightarrow \text{GalNAc} \longrightarrow \qquad 086$$

branches:
$$\begin{array}{cc} \uparrow & \uparrow \\ \text{L Fuc} & \text{Glc} \end{array}$$

$$\longrightarrow \text{GlcNAc} \xrightarrow{1\ \beta\ 2} \text{Glc} \xrightarrow{1\ \alpha\ 4} \text{Gal} \xrightarrow{1\ \alpha} \qquad 0111$$

branches:
$$\begin{array}{cc} \uparrow 6 \ \alpha\ 1 & \uparrow 4 \ \alpha\ 1 \\ \text{Col} & \text{Col} \end{array}$$

Figure 3.18: Neutral O-antigen Oligosaccharide Repeating Units of *E. coli* (Rib, ribose; Fuc, fucose; Col, colitose)

grouping of similar O-side chains into chemotypes. The O-side chains of *Salmonella typhimurium* consist of a tetrasaccharide repeating unit, containing two hexoses, D-galactose and D-mannose; a deoxyhexose, L-rhamnose and a dideoxyhexose, abequose (Figure 3.17). Dideoxy-hexoses are relatively uncommon in nature but are often present in O-side chain polysaccharides. The side chain is attached to the 4 position of the terminal glucose of the LPS core. The O-side chains of many other salmon-ellae also contain galactose, mannose and rhamnose (and in some instances additional glycosyl or acetyl substitutions) often in an identical sequence to *S. typhimurium* (Figure 3.17). However, since the sugars present in the polymer may be linked in either the α or the β configuration they possess distinct antigenic determinants.

Unlike salmonellae, which appear only to possess neutral polysaccharide O-side chains, *E. coli* has been shown to have both neutral and acidic O-antigens. The specific neutral polysaccharide of *E. coli* may be composed of up to six different sugars, often containing unusual amino sugars and rhamnose (Figure 3.18). Only two homopolysaccharide

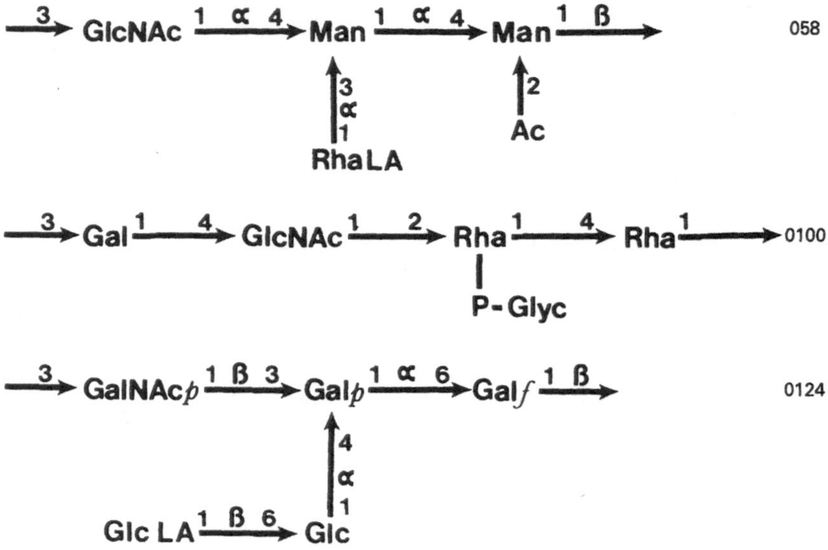

$$\xrightarrow{3} \text{GlcNAc} \xrightarrow[\alpha]{1 \quad 4} \text{Man} \xrightarrow[\alpha]{1 \quad 4} \text{Man} \xrightarrow{1 \quad \beta} \qquad 058$$

$$\begin{array}{c} \uparrow 3 \\ \alpha \\ 1 \\ \text{RhaLA} \end{array} \qquad \begin{array}{c} \uparrow 2 \\ \text{Ac} \end{array}$$

$$\xrightarrow{3} \text{Gal} \xrightarrow{1 \quad 4} \text{GlcNAc} \xrightarrow{1 \quad 2} \text{Rha} \xrightarrow{1 \quad 4} \text{Rha} \xrightarrow{1} \qquad 0100$$

$$\begin{array}{c} | \\ \text{P- Glyc} \end{array}$$

$$\xrightarrow{3} \text{GalNAc}p \xrightarrow{1 \quad \beta \quad 3} \text{Gal}p \xrightarrow{1 \quad \alpha \quad 6} \text{Gal}f \xrightarrow{1 \quad \beta} \qquad 0124$$

$$\begin{array}{c} \uparrow 4 \\ \alpha \\ 1 \end{array}$$

$$\text{Glc LA} \xrightarrow{1 \quad \beta \quad 6} \text{Glc}$$

Figure 3.19: Acidic O-antigen Oligosaccharide Repeating Units of *E. coli* (RhaLA, rhamnolactylic acid [2-*O*-(1'-carboxyethyl)-L-rhamnose] and GlcLA, glucolactylic acid [4-*O*-(l'-carboxyethyl)-D-glucose])

O-specific antigens have been described in *E. coli*, i.e. the 08 and 09 mannans. O-specific mannans are common in *Klebsiella*; the *E. coli* 09 and *Klebsiella* 03 are identical and the *E. coli* 08 and *Klebsiella* 05 are very similar. Examination of the sugar composition of acidic O-specific side chains in *E. coli* has shown them to consist of hexuronic acids, neuraminic acid or 4-*O*-(l'-carboxyethyl)-D-glucose (glucolactylic acid) and other acidic sugars (Figure 3.19).

The composition of the lipopolysaccharide from *Bordetella pertussis*, the causative organism of whooping cough, has received considerable attention. Although the core and O-side chain regions are unremarkable in composition, it appears that this organism is highly unusual in that two polysaccharide chains are attached to the lipid A (c.f. a single chain in other Gram-negative organisms). One of the chains (polysaccharide 1) can be released from the LPS by mild, acid-catalysed hydrolysis and contains a reducing-terminal KDO residue. The second chain (polysaccharide 2) requires more violent hydrolysis conditions to remove it from the LPS, and contains a reducing-terminal KDO-5-phosphate residue.

Biosynthesis of Lipopolysaccharide

Advances in the determination of biosynthetic pathways are greatly facilitated by the isolation of mutants blocked at various points of the assembly sequence and by the use of specific inhibitors. Mutants were invaluable in the elucidation of the structure of the R-core and O-side chain of the LPS. However, this approach could not be used to directly determine the route to the synthesis of the lipid A-KDO region since, as this structure is essential for the structural and functional integrity of the outer membrane, mutants blocked in its biosynthesis would not be viable. Studies of the mechanism of action of antimicrobial agents can give much information concerning cellular biosynthesis; indeed our views of bacterial cell wall synthesis depends greatly on studies using β-lactam and other antibiotics. Unfortunately no specific inhibitors of the early stages of LPS biosynthesis have yet been discovered. Hence our understanding of the assembly sequence of the lipid A-KDO region is rather incomplete.

It has proved possible to isolate temperature-sensitive mutants of *Salmonella typhimurium* which, although growing normally at 30°C, fail to synthesise KDO at higher temperatures, e.g. 42°C. In enteric bacteria KDO is synthesised from ribulose-5-phosphate by three enzymes: D-phospho-arabinose isomerase, KDO-8-phosphate synthetase and KDO-8-phosphate phosphatase (Figure 3.20). The KDO is then converted into the nucleotide sugar cytidine monophosphate-KDO (CMP-KDO), which is capable of donating KDO to the developing lipopolysaccharide. The mutants showing temperature-sensitive KDO-synthesis were shown to be defective in KDO-8-phosphate synthetase at the non-permissive temperature. However, synthesis of LPS precursors continues and as these appear not to be transferred to the outer membrane, they accumulate in the cell. The incomplete lipid A possesses the β1→6-linked glucosamine disaccharide, substituted with 3-hydroxymyristic acid and the two phosphate groups of the complete molecule, but lacks both KDO and the saturated non-hydroxy fatty acids (Figure 3.21). Depending upon strain and cultural conditions, the phosphorylethanolamine and 4-amino-arabinose substituents may be present or absent. However, one of the KDO-deficient temperature sensitive mutants, when grown at the non-permissive temperature (42°) produces the so-called acidic precursor, which lacks the ethanolamine and 4-amino-arabinose moiety, and contains only mono-ester phosphate (Figure 3.22). However at lower growth temperatures these strains produce the neutral precursor form. Another series of temperature-sensitive mutants defective in LPS biosynthesis produces a presumed lipid A precursor termed lipid X, which has been shown to be a diacylated glucosamine-1-phosphate.

Figure 3.23 gives a provisional scheme outlining the biosynthesis of the early stages of LPS biosynthesis. Glucose-1-phosphate is first aminated at C-2 and then 3-OH myristic acid amide linked. The diacyl glucosamine-1-

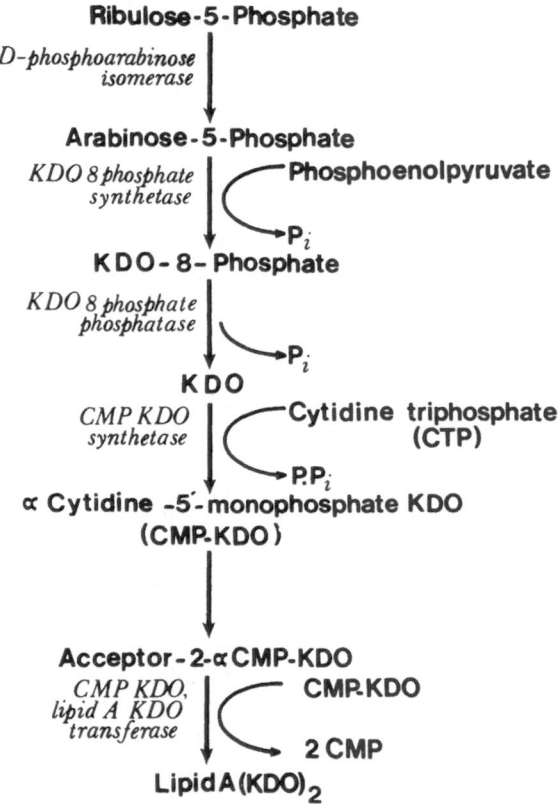

Ribulose-5-Phosphate

D-phosphoarabinose isomerase

Arabinose-5-Phosphate

KDO 8 phosphate synthetase — **Phosphoenolpyruvate**

$\to P_i$

KDO-8-Phosphate

KDO 8 phosphate phosphatase

$\to P_i$

KDO

CMP KDO synthetase — **Cytidine triphosphate (CTP)**

$\to P.P_i$

α Cytidine -5'-monophosphate KDO (CMP-KDO)

Acceptor-2-α CMP-KDO

CMP KDO, lipid A KDO transferase — **CMP-KDO**

$\to$ **2 CMP**

Lipid A (KDO)$_2$

Figure 3.20: Biosynthesis of KDO (3-deoxy-D-*manno*-octulosonic acid) in *Salmonella* and *E. coli*

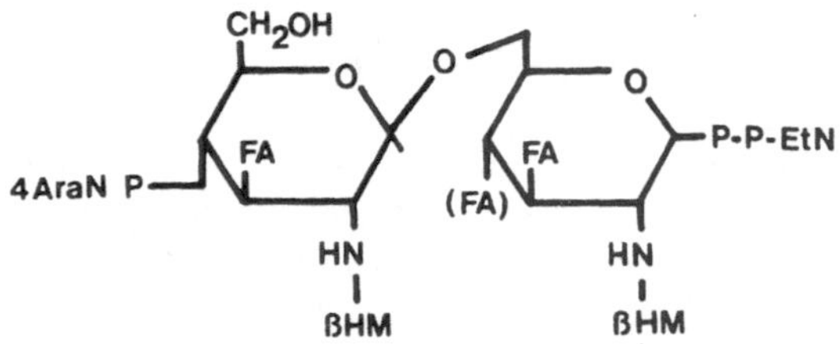

CH$_2$OH

4AraN P— FA   FA —P-P-EtN

(FA)

HN   HN

βHM   βHM

Figure 3.21: Partial Structure of the Neutral Lipid A Precursor

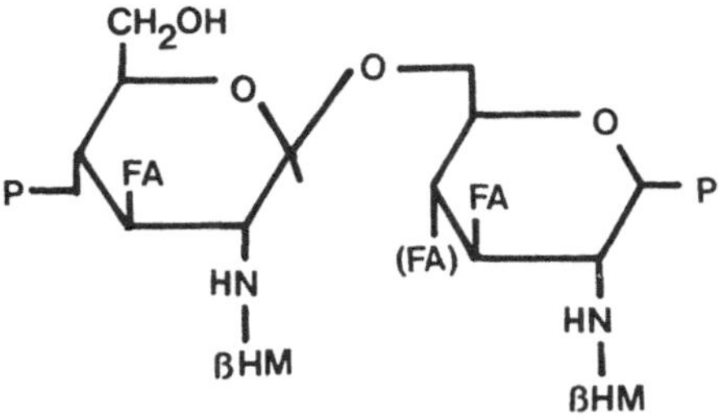

Figure 3.22: Partial Structure of the Acidic Lipid A Precursor

P, presumably equivalent to lipid X, is formed by the addition of another 3-OH fatty acid at C-3. The diacyl glucosamine phosphate is then converted into a nucleotide sugar by reaction with uridine triphosphate, this condenses with diacyl glucosamine which is subsequently phosphorylated to give a structure resembling the acidic precursor produced by KDO-8-P synthetase-deficient mutants. Such a scheme suggests that the neutral and acid precursors of lipid A are direct intermediates in the lipid A biosynthetic pathway, not abortive side products. This has been confirmed by pulse experiments, which have shown that on shifting from non-permissive to permissive temperatures, pulse-labelled precursors are efficiently converted into complete LPS. There is strong evidence to suggest that the acid precursor is the direct acceptor of the optional 4-amino-arabinose and phosphorylethanolamine. Since serine auxotrophs of *Salmonella typhimurium* produce an LPS containing ethanolamine in the presence of serine, but not in its absence, it has been suggested that the phosphorylethanolamine residue derives from serine. This has been confirmed by demonstrating that radiolabelled serine is efficiently incorporated into lipid A.

Incorporation of KDO into the precursor precedes the incorporation of ester-linked straight chain fatty acids. In *S. typhimurium* glycosylation of lipid A precursor is carried out by a membrane-bound enzyme system. Unlike the glucosyl- and galactosyl-transferases involved in the synthesis of the R-core of the LPS, the CMP-KDO transferase system is endowed with a great avidity for the hydrophobic environment of the membrane; indeed it can only be released by treatment at alkaline pH with non-ionic detergents. The detergent-soluble KDO transferase complex is capable of sequentially transferring two moles of KDO from CMP-KDO to the 6′ position of the lipid A precursor. It seems likely that separate transferases are necessary for the addition of the first and second KDO residues.

Little is known concerning the nature of the acyl transferase reaction

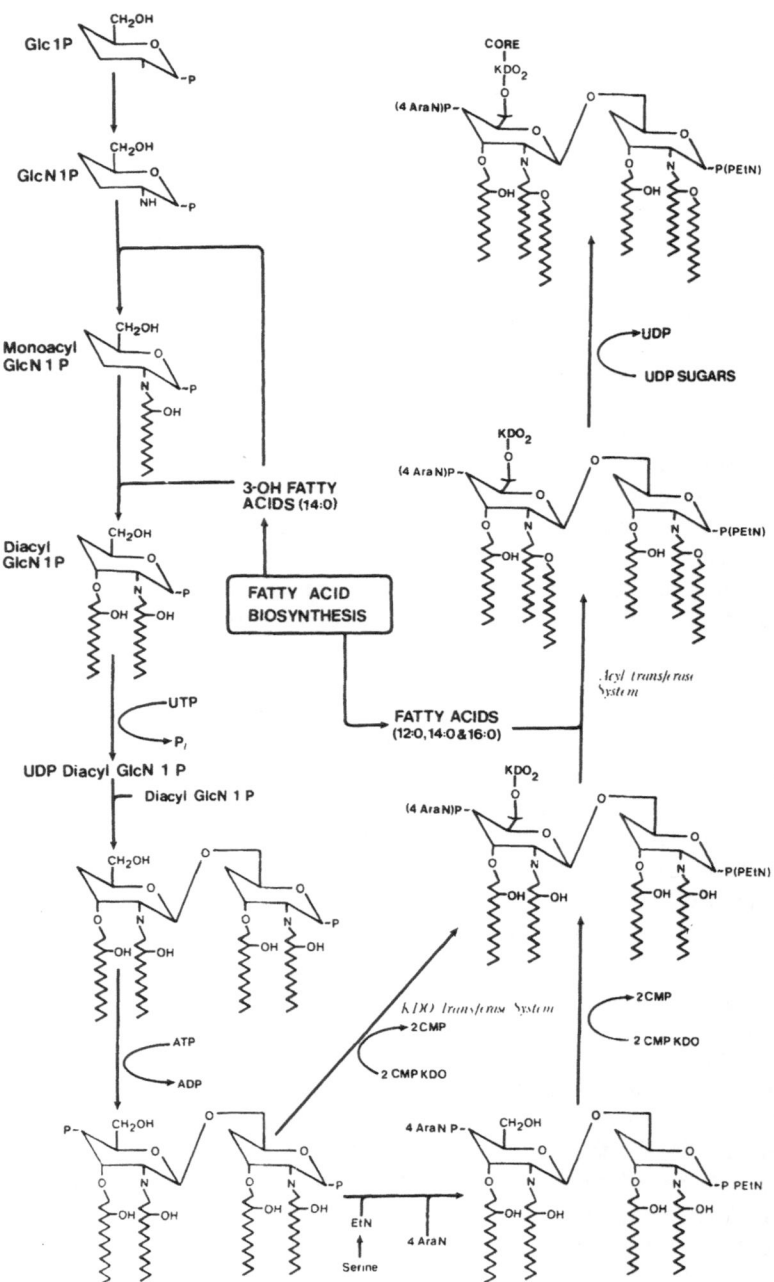

Figure 3.23: A Provisional Scheme for the Biosynthesis of the Lipid A-core Region of *Salmonella*

responsible for the secondary addition of ester-linked fatty acids to lipid A. Deep rough (R_e) mutants of *S. typhimurium* are unable to incorporate the first heptose moiety in the core polysaccharide (Figure 3.16). Examination of the R_e LPS has shown it to contain the full complement of *O*-acyl side chains, suggesting that the final acylation reactions are completed before the polysaccharide chain is extended. However, secondary acylation is not a prerequisite for polysaccharide extension. Addition of the inhibitor of fatty acid biosynthesis cerulenin, prevents the secondary acylation of precursor, but has little effect on chain elongation. Attempts to transfer fatty acids from acyl CoA to the lipid A precursor *in vitro* have not yet been successful. Mutants of *E. coli* unable to synthesise fatty acids can be grown if supplied with exogenous fatty acids. Although under these conditions phospholipid synthesis occurs normally, the LPS does not contain the secondary acyl substituents, suggesting that fatty acyl CoAs cannot act as donors for acylation of lipid A. In bacterial phospholipid synthesis both acyl CoAs and acyl carrier proteins (ACP) can serve as fatty acid donors, and it may be that only acyl ACPs can acetylate the lipid A precursor. The observation that the secondary acylation is not obligatory may provide one explanation of the differences observed between researchers, in the number of acyl constituents reported in purified LPS.

The R-core is sequentially assembled by the transfer of sugars, phosphate and ethanolamine to the lipid A-KDO moiety. Mutants that have lost the ability to synthesise or transport the necessary sugars, or deficient in the production of UDP-sugar derivatives, produce an incomplete core which lacks all sugars distal to the lesion (Figure 3.16). A strain of *Salmonella typhimurium* defective in the enzyme galactose epimerase cannot synthesise UDP-galactose and would produce a R_e-type core lacking the terminal glucose and *N*-acetylglucosamine residues (even in an abundance of UDP-glucose and UDP-*N*-acetylglucosamine).

The biosynthesis of the R-core has been studied in cell-free systems only in enteric bacteria. Although core biosynthesis may be broadly similar in other species, differences are to be expected. The mechanism by which the heptose units are added to the fully acylated lipid A-KDO moiety is obscure. The exact nature of the heptose precursor is not known. Once in place the heptose I sugar is phosphorylated by ATP (Figure 3.24). In *Salmonella* LPS the heptose II unit is also phosphorylated, but this appears to occur after the addition of the first glucose residue. By expeditious use of core mutants and cell-free extracts, followed by biochemical analysis of the product, the sequence of core assembly has been determined for *Salmonella* (Figure 3.24). UDP-sugars interact one at a time with the developing R-core. An appropriate transferase enzyme catalyses each step, binding to the incomplete core acceptor and only dissociating from the complex once the transfer is completed. Phosphorylation of both heptose units is a pre-requisite for the transfer of the first galactose unit to the core.

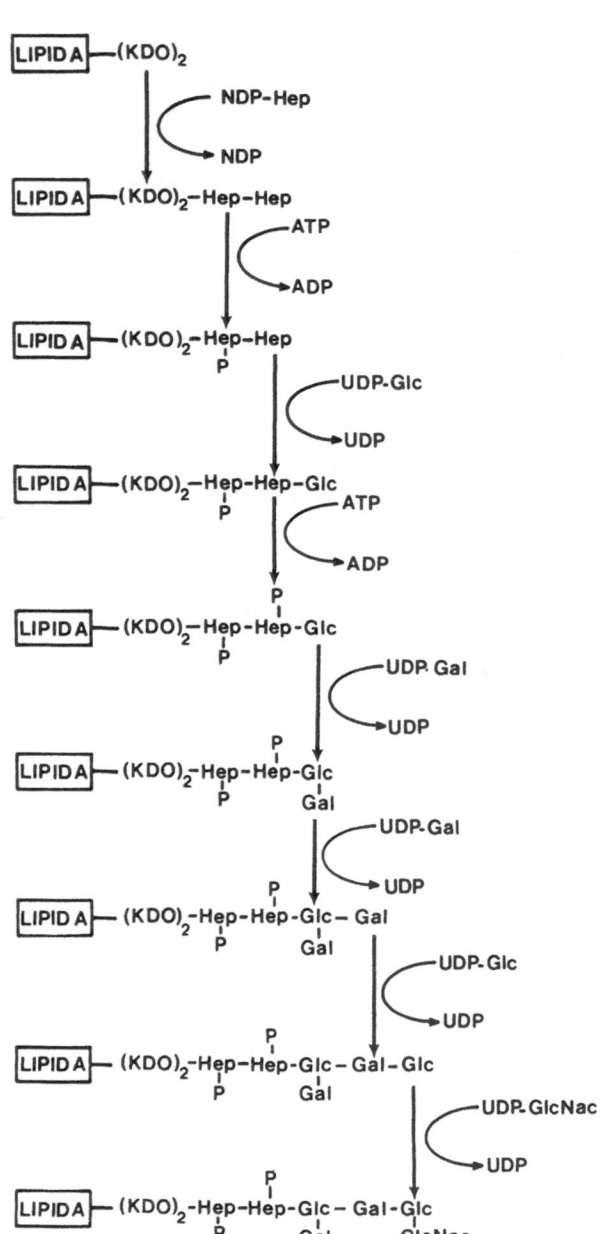

Figure 3.24: Biosynthesis of the Core Region of Lipopolysaccharide in *Salmonella spp.*

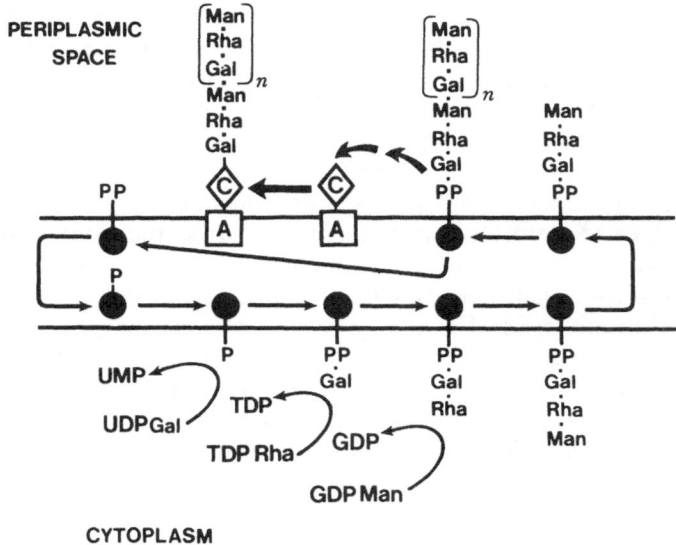

Figure 3.25: Biosynthesis of the O-side Chain Polysaccharide in *Salmonella anatum*. At the inner face of the cytoplasmic membrane the nucleotide sugars are transferred in sequence to the C_{55} polyisoprenoid phosphate carrier lipid (●). Following the transfer of the mannosylrhamnosyl-galactose trisaccharide-carrier lipid to the outer face of the membrane the trisaccharides polymerise forming the lipid-linked O-antigen polymer, the polysaccharide portion of which is then passed to the lipid A: core (A:C). The liberated carrier lipid is then dephosphorylated and returned to the cytoplasmic face of the membrane.

The structural conformation of the developing core appears to determine which transferase enzyme binds to the core, and hence which sugár is next added.

Synthesis of the O-side chain polysaccharide is mediated by a series of cytoplasmic membrane-bound enzymes. The sugar units destined to make up the O-polysaccharide are transferred from sugar nucleotide diphosphates to the C_{55} polyisoprenoid carrier, identical to the lipid involved in peptidoglycan biosynthesis (p. 14). *Salmonella anatum* possesses a relatively simple O-side chain (Figure 3.17) and a scheme for its biosynthesis is shown in Figure 3.25. The general principles of this scheme may also apply to more complex *Salmonella* O-side chains. In *S. anatum* galactose, rhamnose and mannose are sequentially transferred from nucleotide sugar diphosphate to the isoprenoid carrier producing the mannosylrhamnosyl-galactose repeating unit of the O-polysaccharide. The reaction has three distinct steps. Initially, galactose-1-phosphate is transferred from UDP-galactose to produce carrier lipid-P-P-galactose and UMP. The reaction is competitively inhibited by exogenous UMP. The second stage is the transfer of rhamnose from TDP-rhamnose to carrier lipid-P-P-Gal to produce carrier lipid-P-P-galactose rhamnose and subsequent transfer of mannose from GDP-mannose to give the carrier lipid-P-P-galactose, rhamnose, mannose trisaccharide. It should be stressed that all three reactions are catalysed by enzymes located in the cytoplasmic membrane, presumably at the inner face. The trisaccharide lipid carrier must then be translocated to the site of the lipid A-core, which is presumably situated at the outer face of the cytoplasmic membrane. When the trisaccharide-lipid carrier complex reaches the outer membrane the galactosyl-phosphate bond of the trisaccharide-carrier lipid is broken and the galactosyl bond transferred to the terminal mannosyl residue of a second or acceptor lipid forming a hexosaccharide-lipid carrier. This is repeated several times, chain growth occurring in trisaccharide lengths, always to the reducing end of the polysaccharide chain, i.e. the developing polysaccharide is always transferred to an isoprenoid carrier bearing a trisaccharide. A specific phosphatase dephosphorylates the liberated isoprenoid carrier lipid diphosphate to produce the monophosphate derivative, which then returns to the inner cytoplasmic face, available once more to receive another monosaccharide from a nucleotide sugar.

The O-side chain of *S. anatum* is a simple straight chain (with no side chains or substitutions). However examination of the O-side chain polysaccharide of many other *Salmonella* species (Figure 3.17) shows them to consist of branched oligosaccharide, often with additional chemical substitutions. The O-side chain of *S. typhimurium* bears an abequose sugar as a side group. The abequose residue is incorporated into the lipid-linked oligosaccharide before polymerisation into the developing O-side chain, i.e. all the repeating units bear an abequose residue. However in members of the *Salmonella* E_3 serogroup, e.g. *S. minneapolis*, the mannose, rhamnose, galactose backbone of the O-side chain is glucosylated after polymerisation. The substitution of glucose at the C-4 position of the galactose residue is often incomplete, i.e. not all the galactose sugars are substituted. In a similar way the glucosylation of the O-polysaccharide of *S. typhimurium* is carried out after the polymer is formed. It appears that an independent sugar-lipid carrier complex transfers a glucose residue from UDP glucose in the cytoplasm to the O-side chain. Many O-side chains bear acetyl groups. Acetylation occurs during the elongation of the polysaccharide at some stage before the transfer of the sugar polymer to the R-core. Acetylation of the O-side chains of *salmonella* species can be carried out *in vitro*, and acetly-CoA has been shown to be the donor of the acetyl group.

The completed polysaccharide O-side chain, attached to the lipid carrier appears to be freely mobile in the outer leaflet of the cytoplasmic membrane, and at some stage will come into close association with the lipid A-core complex. The transfer of the O-side chain polysaccharide from the lipid carrier to the glucose II of the *Salmonella* core is catalysed by O-antigen LPS ligase. The ligation reaction is independent of the length of the O-side chain; the enzyme seeming incapable of recognising the length of the chains that it transfers. Indeed, mutants unable to carry out the polymerisation of the oligosaccharide repeating unit retain the ability to transfer a single oligosaccharide unit to the lipid A-core complex. Analysis of isolated LPS from smooth strains reveals considerable heterogeneity in the length of the O-side chain polysaccharide. The principle controlling the length of the side chain attained on the carrier before transfer to the core, may be a simple function of the affinity and rate of reaction of the transferase and polymerase enzymes.

A central tenet of the above scheme of O-side chain biosynthesis, is that oligosaccharides linked to lipid carriers traverse the cytoplasmic membrane. There is little direct evidence that this occurs; it being difficult to localise such an intermediate state. Some support for the hypothesis comes from the observation that the antibiotic bacitracin, which specifically interacts with the lipid carrier, inhibits O-side chain polymerisation. However, many consider the transfer of the oligosaccharide from one leaflet of the membrane to the other to pose insurmountable problems. To overcome this difficulty it has been postulated that nucleotide sugars are translocated through the membrane into the periplasmic space, where

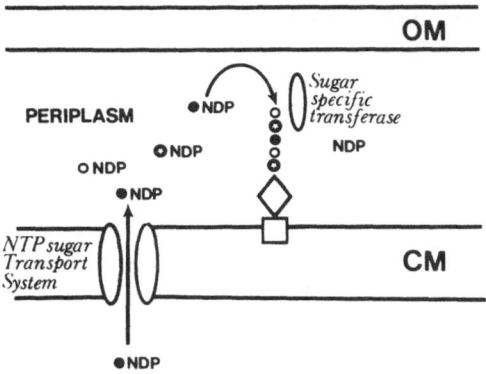

Figure 3.26: Proposed Alternative Mechanism for the Assembly of the LPS O-side Chain. To overcome the necessity for the 'flip-flop' of the O-side chain precursor across the cytoplasmic membrane (see Figure 3.25) it has been suggested that nucleotide sugars could pass through specific transport systems into the periplasm and assembled *in situ* by sugar-specific transferases. There is little direct evidence for such a model

extracellular glycosyl transferases assemble the O-side chain on the outer face of the cytoplasmic membrane (Figure 3.26). Although this hypothesis is attractive thermodynamically, it offers no explanation of why bacitracin inhibits O-side chain elongation. Additionally the presence of the requisite nucleotide sugars in the periplasm has not been established.

The assembly of the LPS, by whatever route, is restricted to the cytoplasmic membrane, yet its ultimate destination is of course the outer membrane. The translocation and subsequent introduction of a molecule as complex as LPS into the outer membrane poses many questions, most of which remain unanswered. The translocation of LPS to the outer membrane is rapid and unidirectional. The residence time of completed LPS in the cytoplasmic membrane is short (with a half time at 30°C of about one minute). Incomplete LPS remains in the cytoplasmic membrane, suggesting that a component of the completed structure, is recognised by the translocation principle. LPS translocation is an active process, being inhibited by uncouplers of oxidative phosphorylation. Completed LPS molecules do not insert in a random fashion over the bacterial outer membrane surface, but rather at a limited number of distinct sites. These sites are thought to correspond to the adhesion points between cytoplasmic and outer membrane. It is not known whether there is a corresponding segregation of the sites of LPS biosynthesis beneath these points or whether LPS assembled over the entire cytoplasmic surface gravitates to these zones. At the adhesion zone, the two membranes are presumably held in close proximity to facilitate the transfer of newly synthesised outer membrane components. A yet uncharacterised system is responsible for the observed rapid unidirectional movement of LPS from the cytoplasmic membrane to the inner face of the outer membrane. Since LPS is only found in the exterior leaflet of the outer membrane a second intramembrane translocation event (i.e. flip-flop) must also occur. LPS shows strong affinity for the protein component of the outer membrane, this interaction being extensive and of high affinity. It is believed that protein-LPS interaction is responsible for the functional irreversibility of LPS translocation to the outer membrane.

Proteins of the Outer Membrane

Membrane proteins can be solubilised by treatment with hot detergent. If sodium dodecyl sulphate (SDS) is used the proteins become coated with a net uniform negative charge. When applied to SDS-polyacrylamide gels, such proteins will migrate towards the cathode, at a rate proportional to the logarithm of their relative molecular mass. The proteins can then be stained with a protein specific dye. SDS-polyacrylamide gel electrophoresis has proved a powerful tool in determining the protein composition of isolated inner and outer membranes of Gram-negative bacteria. In general the

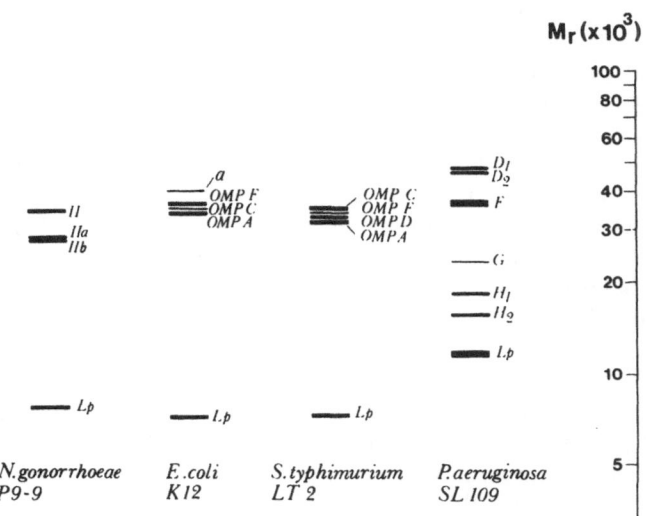

Figure 3.27: SDS-polyacrylamide Gel Electrophoresis Patterns of the Major Outer Membrane Proteins of Four Gram-negative Species. The isolated membrane fractions have been heated to 100°C in SDS before electrophoresis

protein profile of outer membrane is less complex than that of the cyto-plasmic membrane. The cytoplasmic membrane contains between 40 and 100 distinct protein bands, whereas the outer membrane contains a more limited number (Figure 3.27), presumably reflecting the more specialised nature of the membrane. It was at first thought that in the outer membrane of *E. coli* a single protein (M_r 40,000), termed major outer membrane protein, constituted 75 per cent of the total. As the techniques of protein separation improved, this major band has been resolved into at least four distinguishable proteins. This heterogeneity was discovered by several researchers, who unfortunately introduced a series of different nomen-clatures. This confusion is slowly being resolved, and so in *E. coli* and *S. typhimurium*, which have received most study, a standard nomenclature is now possible, based upon the genetic loci controlling the production of the protein (designed *ompA-F*). When running polyacrylamide gels an indicator dye, usually bromophenol blue, is added to the membrane frac-tion to allow the operator to know when low-molecular-weight material reaches the end of the gel. Just behind the dye front, is found a band that reacts poorly on subsequent treatment with protein-specific dyes, e.g. Coomassie blue. Analysis of this band shows it to contain protein plus appreciable quantities of lipid (which presumably accounts for the poor affinity of the protein for the dye). This low-molecular-weight lipoprotein, first described by Braun, in *E. coli* has also been detected in *Salmonella, Proteus, Pseudomonas* and *Serratia* species. In addition to the major outer membrane proteins and lipoproteins, the outer membrane also contains other proteins, the so-called minor proteins. Many of the minor proteins are inducible or derepressible proteins, and may under appropriate environmental conditions reach levels in the outer membrane comparable with those of the major proteins. The outer membrane should therefore not be considered as a structure of invariable composition. The loss of, or changes in the major components can often be accommodated, with little obvious change in the physiology of the organism. This suggests a degree of functional redundancy and cooperativity between constituent proteins.

Lipoprotein

In terms of the number of molecules present, the lipoprotein is the most abundant protein in the *E. coli* cell envelope. It is present in about 7.5 × 10^5 copies/cell, representing about 6 per cent of total cell protein. It occurs in two forms; about a third of the lipoprotein is covalently bound to the peptidoglycan, the remainder existing freely in the outer membrane. The lipoprotein of *E. coli* is an unusual polypeptide, consisting of 58 amino acids, bearing cysteine at the *N*-terminus and a lysine residue at the *C*-terminus. The *N*-terminal residue is covalently linked to a diacyl glycerol by a thioesther bond. The fatty acid constitutes appear unremarkable, appearing similar to that found in phospholipids of the same cells. In the peptidoglycan-linked form the *C*-terminal lysine of the lipoprotein is covalently linked through the ε-amino group to the L-carboxyl group of diaminopimelic acid, linked to a muramic residue of peptidoglycan (Figure 3.28). The non-peptidoglycan-bound lipoprotein contains a free lysine residue at the *C*-terminal. Lipoprotein may be released from peptidoglycan by treatment with enzymes. Trypsin specifically cleaves between lysine (55) and tyrosine (56) and between arginine (57) and lysine (58) (Figure 3.28) releasing the lipoprotein. Alternatively, the peptidoglycan can be dissolved by lysozyme, to free the lipoprotein. Biophysical analysis of isolated lipoprotein reveals two distinct regions of high α-helical content (residues 4 to 25 and 29 to 47) linked by a β-loop. Although Figure 3.28 shows the lipoprotein as a straight rod, it is probable that the two regions of α-helix loop back on each other.

Unlike most of the other outer membrane proteins, the lipoprotein does not serve as a binding site for any bacteriophages; neither can antibodies prepared against isolated lipoprotein, be shown to bind to intact cells. These observations suggest that the proteinaceous portion of the molecule is deeply buried in the envelope. Treatment of *E. coli* cells with trypsin leads to a detachment of the outer membrane from the peptidoglycan,

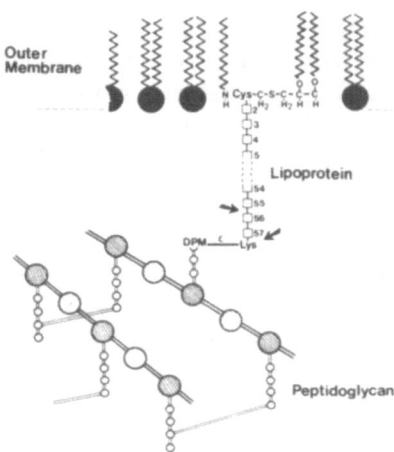

Figure 3.28: Diagrammatic Representation of the Insertion of Lipoprotein into Outer Membrane. The diagram shows a bound-lipoprotein molecule linked to the diaminopimelic acid residue (DPM) of peptidoglycan by its terminal lysine. In free-lipoprotein this terminal lysine would be unsubstituted. Bound lipoprotein can be liberated by trypsin, which cleaves the polypeptide at points marked by arrows. (The symbols used for peptidoglycan structure are as in Figure 1.2.)

suggesting that one function of the peptidoglycan-linked lipoprotein is to anchor the outer membrane to the cell wall. This has been confirmed by examining the properties of *E. coli* mutants defective in lipoprotein biosynthesis. Mutants that completely lack both free and bound lipoprotein (*lpo*-mutants) grow and divide normally. However the mutants are hypersensitive to EDTA and detergents and tend to leak periplasmic proteins into the environment. Examination of electron micrographs of *lpo* mutants reveals many surface vesicles or 'blebs', presumably the result of decreased outer membrane stability in the absence of lipoprotein.

Major Outer Membrane Proteins

The term major outer membrane protein refers to components present in sufficient quantity to be readily detected by simple staining of polyacrylamide electrophoretograms. The protein profiles produced show great pleiotrophy; differences often occurring between strains of the same species, and even in the same strain when grown, say at a different temperature or under nutrient limitation. Proteins located at the cell surface often serve as receptors for bactericidal proteinaceous antibiotics called bacteriocins and bacteriophages. Bacteriophage and bacteriocin-resistant strains and those strains containing lysogenic phage often show modified outer membrane profiles. Identification and characterisation of the major outer membrane proteins has been carried out simultaneously by many researchers, using a variety of membrane isolation techniques, differing detergent-polyacrylamide electrophoresis systems and an enormous range of different strains, resistant to a variety of bacteriophages and bacteriocins. The unavoidable result of this endeavour has been a multiplicity of nomenclatures. To avoid unnecessary confusion of the reader, the structure and function of the major outer membrane proteins is presented in the light of current nomenclature with no reference to the past.

Major Outer Membrane Proteins of E. coli. A notable feature of polyacrylamide gels of *E. coli* K12 outer membrane proteins is a group of proteins in the molecular weight range 35,000 to 38,000 (Figure 3.27). Other *E. coli* strains also produce polypeptides of slightly different molecular weights, but within the same range. The numbers of such bands may vary between one and four. Two types of protein can be identified within this group, those extractable from isolated membrane by 2 per cent SDS at 60°C and those which are not. The major outer membrane protein of *E. coli* BE, a polypeptide of molecular weight 36,500, present in 10^5 copies/cell, can only be extracted at the higher temperature. The protein is tightly associated with the peptidoglycan. High-resolution electron microscopy shows this protein to be arranged on the outer face of the peptidoglycan in a lattice-like structure possessing hexagonal symmetry. This symmetrical arrangement persists even after solubilisation of the peptidoglycan by treatment with lysozyme. The ability of this protein to form stable symmetrical configurations prompted the name matrix protein. Proteins similar to the *E. coli* BE matrix protein are also present in K12 strains of *E. coli.* The OmpC and OmpF matrix proteins of *E. coli* K12 are tightly bound to the peptidoglycan layer. Since they are also known to serve as receptors for bacteriophages and bacteriocins they must completely span the outer membrane (Figure 3.2) and serve as non-specific channels through the membrane, hence their alternative name 'porins'. Unlike the matrix proteins the OmpA protein, of *E. coli* K12 outer membrane is readily extractable with 2 per cent SDS at 60°C. The OmpA protein, like most other readily extractable major outer membrane proteins, exhibits apparent changes in molecular weight (as determined by SDS-polyacrylamide gel electrophoresis) when heated during sample preparation and has been termed a 'heat modifiable' protein. Our understanding of the structure, topography and function of major outer membrane proteins has been aided by the use of mutants resistant to bacteriophages and bacteriocins which bind to specific outer membrane components (Table 3.2). Mutants of *E. coli* K12 lacking OmpF, are characterised by tolerance to bacteriocin JF246 and resistance to the Tula phage. Similarly mutants resistant to the phage PA2 or Me1 have been shown not to produce the OmpC. A third

Table 3.2: Properties of the Constitutive Major Outer Membrane Proteins of *Escherichia coli* K12

Outer membrane component	Relative molecular mass	Number of molecules per cell	Phage or colicin receptor	Function
Lipoprotein	7,200	bound: 2.5×10^5 free: 5×10^5	—	Stabilises the outer membrane by anchoring outer membrane to peptidoglycan
OmpA	35,160	10^5	K3, Tu II*	Role in conjugation; stabilises mating bacteria
OmpC	36,000	10^5	PA2 Tu Ib T4	Porin, general pore for hydrophilic solutes of $M_r > 700$
OmpF	37,200	10^6	Tu Ia, T2	Porin, general pore for hydrophilic solutes of $M_r > 700$
Protein *a*	40,000	4×10^4	LP81	Protease; regulation of capsular biosynthesis

class of mutants, which are resistant to OmpF and OmpC specific phages and bacteriocins, maps at a separate locus in the *E. coli* chromosome designated *ompB*. Such mutants lack both the OmpF and OmpC proteins. Genetic studies indicate that *ompF* and *ompC* are the structural genes and *ompB* is the regulatory locus controlling the expression of the two structural gene products. Mutants deficient in both the ompC and ompF proteins exhibit a phenotype with generalised deficiencies in solute translocation. This phenotype can be suppressed by additional mutations that specify the production of distinct new outer membrane protein. Three such suppressor loci, *nmpA, nmpB* and *nmpC* have been described.

Although the outer membrane of *E. coli* is freely permeable to hydrophilic material of molecular weight of 700 or less, higher-molecular-weight compounds are excluded. Reconstituted lipid vesicles can be prepared from phospholipid, LPS and isolated outer membrane proteins. It can be demonstrated that the OmpF and OmpC proteins, when introduced into the lipid vesicle, produce non-specific pores or channels through the artificial lipid bilayers, freely permeable to sugar polymers of molecular weight of 600 or less. Although the OmpF and OmpC porins are not noticeably hydrophobic in terms of their amino acid composition, the native protein shows a great affinity for hydrophobic environments. Upon detergent extraction, they exist as oligomers, which sedimentation equilibrium studies suggest are made up of stable trimers. It is suggested that a porin trimer inserts into the outer membrane while being covalently linked to the peptidoglycan layer (Figure 3.29). Since the porin also serves as a phage and bacteriocin receptor it must also be exposed at the cell surface. It is presumed that the trimer constitutes a general hydrophilic aqueous

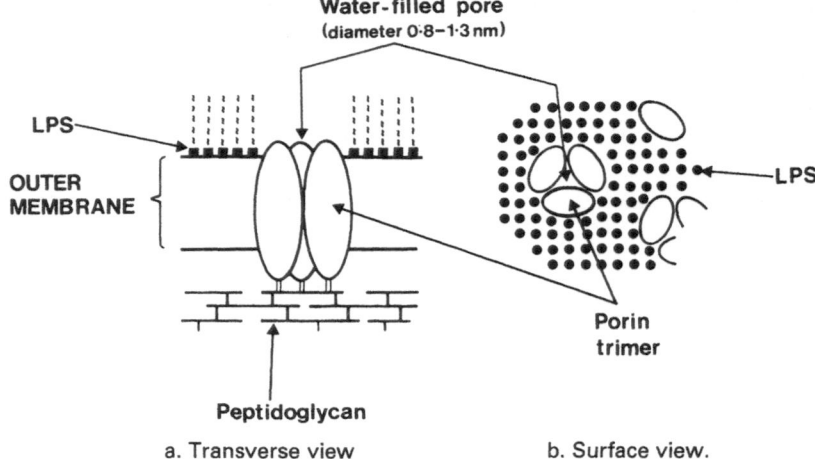

Water-filled pore
(diameter 0·8–1·3 nm)

LPS

OUTER
MEMBRANE

LPS

Porin
trimer

Peptidoglycan

a. Transverse view   b. Surface view.

Figure 3.29: Generalised Structure of Outer Membrane Porins. Although there is strong evidence that the porins exist in a trimeric configuration, the actual pore may exist as shown or be present in each individual porin molecule, i.e. 3 pores/porin trimer

channel that allows passive non-specific diffusion of low-molecular-weight materials across the outer membrane. Biophysical measurements of single-channel conductance and permeability to large organic ions by OmpF proteins integrated into planar lipid films indicate a minimum pore diameter of 0.8nm. Detailed study of the conductance properties of porins in the presence of applied electrical voltages, indicates that OmpF pores exist in two states, open and closed and that the opening of the OmpF channel is voltage-induced and voltage-controlled.

The OmpA major outer membrane protein differs from the porins in several respects. Firstly it is readily extracted from the envelope of *E. coli.* Secondly it exhibits heat modifiable rates of migration on SDS-polyacrylamide gel electrophoresis. Finally OmpA is the only major polypeptide that is sensitive to protease cleavage in isolated envelope fractions. The OmpA protein of *E. coli* acts as a receptor for phages K3 and Tu II and is therefore considered to be exposed at the envelope surface. The exact role of OmpA protein in the physiology of the outer membrane is not yet fully understood: *ompA* mutants, defective in OmpA biosynthesis are unable to act as recipients in conjugation with donor cells bearing F-type pili. Isolated OmpA protein complexed with LPS is able to block F-pilus mediated conjugation. OmpA protein may function directly in promoting the establishment or maintenance of mating-pairs. The OmpA protein is also implicated in the maintenance of the integrity of the outer membrane and cell shape. Although single *ompA* or *lpp* (lipoprotein-less) mutants of *E. coli* are rod shaped, double *lpp, ompA* mutants grow as cocci; suggesting a role for both lipoprotein and OmpA in the determination of cell shape.

The largest major outer membrane protein, designated protein *a*, is present in appreciable quantity in *E. coli* K12 only when the organism is grown at 37°C or above. Protein *a* is able to act as a protease catalysing the *in vivo* cleavage of the outer membrane receptor for ferric enterobactin (p. 94). An additional role for protein *a* in the regulation of capsular polysaccharide formation has also been proposed.

Major Outer Membrane Proteins of Salmonella typhimurium. Analysis of isolated outer membrane of *S. typhimurium* using SDS-polyacrylamide gel electrophoresis reveals the presence of four polypeptides (M_r 33,000 to 36,000; see Figure 3.27) of which three OmpD, OmpF and OmpC (M_r 34,000, 35,000 and 36,000 respectively) have been identified as porins. The *ompD* locus produces a polypeptide of which no counterpart exists in *E. coli.* The OmpC, OmpD and OmpF proteins of *S. typhimurium* are linked to the peptidoglycan. Phage resistance can be used to isolate mutants lacking OmpD and/or OmpC; mutants lacking OmpD may be selected with the phage PH51 and without OmpC using PH105 and P221. Genetic analysis suggests that the 35,000 molecular weight polypeptide of *S. typhimurium* is homologous to the OmpF protein of *E. coli. Salmonella* mutants defective in OmpD and OmpC exhibit a marked reduction in the rate of solute permeation into the periplasm. Isolated porins from *S. typhimurium* are capable of forming hydrophilic pores in artificial bilayers, with comparable properties to those of *E. coli* porins. The expression of *OmpC, OmpD* and *OmpF* is regulated by an *ompB* locus similar to that present in *E. coli.* The smallest major outer membrane protein of *S. typhimurium* (M_r 33,000) is heat modifiable and is analogous to the OmpA protein of *E. coli.* Although phages using the OmpA protein of *S. typhimurium* have not been detected, the protein must have a region at the cell surface since it serves as a receptor for bacteriocin 4–59.

Major Outer Membrane Proteins of Pseudomonas aeruginosa. Members of the genus *Pseudomonas* are renowned for their resistance to a wide range of antibiotics. The drug resistance is thought to be related to the restrictive permeability properties of the outer membrane, since the organism shows increased antibiotic sensitivity upon removal of the outer layers of the envelope by physical or chemical means. The belief that the drug-resistance properties of *P. aeruginosa* derive from the ability of otherwise potent antibacterials to reach effective concentrations at their site(s) of action, has prompted many studies into the permeability properties of the pseudomonad outer membrane. In enteric bacteria material of a molecular weight greater than 700 is excluded by the porins. However, in *P. aeruginosa* significant levels of saccharides, which would be excluded by the outer membrane of enteric bacteria, freely diffuse into plasmolysed *P. aeruginosa* cells. Isolated outer membrane preparations from this organism were freely permeable to saccharides of under 4,000 molecular weight. Studies using isolated porins from *P. aeruginosa* in reconstituted artificial lipid bilayers have confirmed that the exclusion limit of the pore is about 4,000. It has been suggested that the larger pores may be advantageous to organisms that are both proteolytic and lipolytic, since they would permit the uptake of micellar lipids and large peptides by the organism, presumably to be subsequently degraded in the periplasmic space. The observation that the outer membrane of *P. aeruginosa* is freely permeable to hydrophilic solutes of under 4,000 molecular weight is difficult to reconcile with the belief that the drug resistance of the genus derives from restrictive envelope permeability. It may be that the porins have some degree of selectivity, being able to prevent the diffusion of unwanted agents, while permitting the uptake of nutrients.

Figure 3.27 shows the protein profile of isolated outer membrane of *P. aeruginosa*. Four of the bands are heat-modifiable, designated D, F, G and H. Protein D can be resolved into two bands D_1 and D_2, if the polyacrylamide gel strength is greater than 14 per cent. The D and E proteins are strongly induced in minimal medium with glucose as the sole carbon source, but are less evident in membranes isolated from cells grown in a nutrient-rich medium. It is likely that these proteins are involved in the passage of glucose through the outer membrane. Isolated protein D has been shown to act as a glucose-specific pore in reconstituted membranes, and may be analogous to the lamB pore of *E. coli* (p. 92). Protein F of *P. aeruginosa* is the non-specific porin, comparable with the *omp*F product of enteric bacteria. It differs from OmpF in that it is heat-modifiable, although the significance of this is not understood.

Major Outer Membrane Proteins of Neisseria gonorrhoeae. The surface protein antigens of *N. gonorrhoeae* are of great interest to researchers, since they have a role in virulence of the strain and have potential in the production of vaccines against the disease. Considerable variation exists in the pattern of *Neisseria* outer membrane proteins seen in different strains of the organism, but in general the major outer membrane protein (Protein I; M_r 36,000) is accompanied by one or two additional proteins (proteins II) in the molecular weight range 24,000 to 36,000. Protein I spans the outer membrane of *Neisseria* and serves as a hydrophilic diffusion pore in a manner analogous to OmpF of *E. coli*. Strains of *Neisseria* which have developed drug-resistance by modifying their envelope permeability have been shown to possess protein I of increased molecular weight. Workers have noted a series of antigenically distinct proteins I amongst different clinical isolates. The heat-modifiable proteins II of different strains also show antigenic diversity. At least five protein IIs have been identified in *Neisseria* although each strain only seems to possess two protein II bands.

Table 3.3: Minor Outer Membrane Proteins of *E. coli* K12

Outer membrane protein	Relative molecular mass	Conditions for induction or repression	Phage or colicin receptor	Function
Tsx	26,000	(Constitutive)	T6 Colicin K	Pore for nucleosides
PhoE	40,000	Phosphate limitation	TC23 TC45	General pore for hydrophilic solutes; prefers anionic solutes in particular polyphosphates
LamB	50,000	Presence of maltose	λ, K10	Pore for small hydrophilic solutes recognised by maltose and maltodextrins
BtuB	60,000	Vitamin B12 limitation	BF23, E-group colicins	Uptake of vitamin B12
Cir	74,000	⎫	Colicins I and V	Fe^{3+}-uptake
TonA	78,000	⎪	T1, T5, Φ 80 Colicin M	Ferrichrome uptake
FecA	80,500	⎬ Iron limitation	—	Ferric citrate uptake
FepA	81,000	⎪	Colicin B	Fe^{3+}-enterobactin uptake
83K protein	83,000	⎭	—	Uptake of complexed iron

Unlike the type I protein, these proteins appear to be surface located. Protein IIa may be involved in the interaction of the bacterium with leucocytes.

Minor Outer Membrane Proteins

With the development of improved methods of outer membrane separation and the use of higher resolution SDS-polyacrylamide gel electrophoresis it became clear that some of the minor bands present in the gels were not cytoplasmic contaminants but *bona fide* outer membrane components. These proteins have a diversity of structure and function, their only common feature being their presence in the outer membrane in low copy number. Certain of the minor proteins are constitutive but many of the others are inducible or derepressible. Under the appropriate environmental conditions the production of these minor proteins is greatly increased and the proteins may be made in quantities approaching that of the major outer membrane proteins. Many of the minor proteins facilitate specific diffusion processes, usually of essential solutes that are too large to use the general porins. Discussion of the properties of minor outer membrane proteins will, in the main, be confined to *E. coli*, the only species for which a coherent account can currently be given (Table 3.3).

Mutants of *E. coli* defective in the *tsx* locus are resistant to bacterio-phage T6 and colicin K (bacteriocins produced by one strain of *E. coli* lethal to other strains of the same organism are termed colicins). The phage- and colicin-resistant mutants are defective in nucleoside uptake, even though the *omp* genes are fully expressed. The protein is thought to function as a pore rather than a carrier because nucleoside substrates do not appear to compete with each other for uptake or inhibit T6 absorption. The necessity for a nucleoside-specific pore in the outer membrane is not obvious, since the molecular weights of the nucleosides are sufficiently low to allow them to use the general OmpC and OmpF porins.

Phosphate limitation of *E. coli* K12 cultures results in a derepression of a protein, PhoE, whose apparent function is to scavenge trace amounts of phosphorous-containing nutrients from this environment. The PhoE protein is immunologically related to OmpC and OmpF and seems to be produced at their expense under phosphate-limitation. The PhoE porin exhibits a preference for anionic solutes, particularly pyrophosphates. The protein bears a specific phosphate binding site. The PhoE protein facilitates the growth of the organism on polyphosphate, provided as the sole source of phosphorus.

The outer membrane receptor for the λ bacteriophage is specified by the *lamB* gene, which is part of the maltose operon. Synthesis of the LamB protein is controlled by the maltose regulatory gene *malT* and is a protein of molecular weight 50,000, present exclusively in the outer membrane. Mutants defective in the *lamB* gene exhibit reduced maltose uptake, while accumulation of maltotriose is reduced to zero. However, isolated LamB protein does not show any affinity for maltose, nor does it protect the bacterium against infection by λ phage. Growth of *E. coli* in the presence of maltose derepresses the LamB protein. The LamB protein forms a pore in the outer membrane, which can be used by a variety of solutes, but has a predilection for maltose. Since maltose itself is small enough to pass through the general porins, the main role of the LamB protein pore is the specific transport of maltodextrins which are too large to use the general porin route. Detailed study of the LamB pore suggests that it is not simply the Stokes radius of a molecule that determines the permeation of solutes through the pore. The LamB pore can mediate the uptake of maltodextrins up to maltoheptaose (seven sugar residues) and act as a non-specific pore for many monosaccharides. However, other simple solutes, e.g. histidine and 6-aminopenicillanic acid are unable to use the LamB pore, suggesting that the pore can discriminate between solutes. The LamB protein is believed to exist as a peptidoglycan-associated trimeric pore in the outer membrane. Biophysical determination of the effective diameter of the LamB pore inserted into artificial planar lipid bilayers suggests that the pore formed by LamB is larger than that of the OmpF channel (1.3 nm against 0.8 nm). The accumulation of maltose by *E. coli* is dependent upon the presence of a periplasmic maltose binding protein. The maltodextrins bound to the binding protein are subsequently actively transported across the cytoplasmic membrane, a process that is mediated by several specific carrier proteins. These carriers are induced simultaneously with the LamB protein. A maltose induced protein (M_r 44,000), analogous to LamB, is present in the outer membrane of *S. typhimurium*. The glucose-induced proteins of *Pseudomonas aeruginosa* are similar in many respects to LamB.

E. *coli* is unable to synthesise vitamin B12 and does not normally require it for growth, although exogenous vitamin is readily transported into the cell. Mutation within the *btuB* gene impairs the uptake of vitamin B12. The *btuB* gene codes for a 60,000 molecular weight outer membrane receptor that requires divalent cations and LPS for maximum activity. The BtuB protein is present in low numbers (about 250 copies/cell) and binds vitamin B12 with a high affinity (Kd for B12 = 0.3×10^{-4}M). The initial rapid binding of the vitamin to the receptor is followed by an energy-dependent transfer of the vitamin to the cell interior. A specific transport event is essential for the accumulation of vitamin B12, since its molecular weight (1355) precludes its passage through the general porins.

The BtuB protein also serves as a receptor for E-group colicins and the phage BF23. Although sharing a common surface receptor different uptake transport mechanisms are involved in the subsequent translocation step into the cell. Mutation of the *btuA* gene impairs B12 transport without affecting sensitivity to E colicins or BF23. Specific inhibition of BtuB synthesis in *E. coli* results in the rapid loss of sensitivity to colicins E2 and E3, followed later by resistance to the BF23 phage. Vitamin B12 uptake continues under these conditions. This suggests that there are different functional states of the BtuB protein and that only newly synthesised protein is functional as a receptor for colicins and phage. It is possible that the newly synthesised protein is orientated in such a way as to facilitate the lethal absorption of phage or bacteriocin. Subsequently, changes in the conformation of the envelope lead to loss of the preferred orientation, firstly in respect to colicin uptake and later with respect to phage binding. Vitamin B12 uptake appears not to be affected by changes in the orientation of BtuB. It may be that newly synthesised BtuB protein emerges on the cell surface at adhesion sites, and it is only at these sites that it is an effective receptor for phage and colicin. As the protein diffuses from the adhesion site it cannot bind phage or colicin but retains its affinity for the vitamin.

Iron is a vital constituent of many bacterial enzyme-systems. Most micro-organisms have evolved high-affinity pathways for the assimilation of Fe(III), which utilise low-molecular-weight carriers or siderophores (iron-bearers). A carrier is essential to iron accumulation since very little of it is available in aerobic environments. The problem is not in the main caused by low exogenous iron levels, indeed iron is the fourth most abundant element on the surface of the earth, but one of availability. In many situations it is a consequence of the near insolubility of Fe(III) at biological pH (Ksol Fe(OH)$_3$ = 10^{-38}M) and of the formation of insoluble oxyhydroxide polymers. This limits the solubility of Fe(III) at neutral pH to about 10^{-18}M. The problem is compounded in organisms growing in a vertebrate host, where the bacterium is in direct competition with the host for available iron. Within body fluids the amount of free iron, which might be readily available to bacteria, is extremely small. This is due to the presence of high affinity iron binding proteins, transferrin in blood and lymph, and lactoferrin in mucosal secretions and milk. Iron is very tightly bound by these proteins; the association constant for lactoferrin has been estimated as 10^{36}, producing free iron concentrations lower than 10^{-18}M, a level too low to support maximal growth rates. For any pathogenic organism to be able to grow in such iron-restricted situations it must be endowed with the ability to sequester iron, in the face of severe competition from the host's iron-binding proteins. Siderophores produced by the bacteria complex with any free iron and in some cases can even take iron from transferrin or lactoferrin. The resulting iron-siderophore complex is too large to utilise the general porin channels necessitating the provision of specific iron-transport systems on the bacterial cell surface.

Under conditions of iron starvation enteric bacteria excrete large quantities of the siderophore enterobactin (also known as enterochelin). Enterobactin is capable of solubilising Fe(III) that would otherwise be unavailable to the cell. The association constant of enterobactin has been estimated as 10^{52}. Since the enterobactin: Fe(III) complex diffuses through the outer membrane it must be transported into the cell by a specific high affinity system. Growth of *E. coli* or *S. typhimurium* in iron-deficient medium induces the production of five outer membrane proteins (Table 3.3). The 81,000 molecular weight protein appears to be the binding site of enterobactin-Fe (III) and colicin B. Mutation of the *fepA* gene leads to the loss of this polypeptide. Freshly isolated outer membrane preparations bind enterobactin, but on incubation of the membranes at 37°C the binding potency declines. This event can be correlated with processing of the FepA protein to a new product 81K*, which appears to have lost a 6,000 M$_r$ polypeptide. This processing of FepA is mediated by the major outer membrane protein, protein *a* (p. 89). The 81K* polypeptide appears unable to bind enterobactin-Fe(III) or colicin B.

A second high-affinity iron-transport system is also present in enteric bacteria, mediating the accumulation of iron complexed with ferrichrome. Mutants of *E. coli* defective in the T1 phage receptor (*tonA*, recently redisignated *fhuA*) are incapable of ferrichrome transport and fail to produce a 78,000 molecular weight outer membrane protein. The TonA protein serves as a receptor for phages T1, T5, Φ80 and colicin M; the isolated protein inhibiting the lethal action of the phages and bacteriocin. Ferrichrome is able specifically to prevent the absorption of colicin M and Φ80 to sensitive bacteria. It should be stressed that the *tonA* iron uptake system is quite independent of the enterobactin uptake system. A similar protein to TonA occurs in the outer membrane of *S. typhimurium* LT-2 and serves as the absorption site for phage ES18.

A third high-affinity iron-uptake system mediates the accumulation of iron complexed with citrate. Growth of *E. coli* in the presence of citrate induces the production of an outer membrane protein (M_r 80,500). Genetic analysis of the iron citrate transport system of *E. coli* located the genes controlling the system, designated *fec*. Subsequently the *fec* locus was subdivided into *fecA* and *fecB*, the former coding for the outer membrane protein.

A 74,000 molecular weight outer membrane protein, designated Cir is also regulated by iron concentration in *E. coli* K12. This protein is the receptor for I colicins and is absent in I-resistant strains. The exact role of this component, and of the 84K iron-induced outer membrane polypeptide, in Fe(III) transport is not clear.

The functional locus *tonB* plays a central role in the processing of ligands bound to the Cir, TonA, FecA, FepA and BtuB outer membrane proteins of *E. coli*. Mutation in the *tonB* locus abolishes high affinity transport by any of the iron-specific membrane proteins and additionally prevents the lethal action of colicins B, D, I, M and V or infection by T1 and Φ80 phage. In *tonB* mutants the number of iron receptors present in the membrane is greatly increased suggesting that a step subsequent to ligand binding is involved. The uptake of enterobactin-Fe(III) is strongly inhibited by uncouplers of oxidative phosphorylation. Such agents dissipate the energised state of the cytoplasmic membrane. The permeability of the enterobactin-Fe(III) transport system appears to be directly controlled by the energy state of the cytoplasmic membrane and *TonB* serves as a device to couple the two membranes in some way. Similarly the energy-dependent uptake of vitamin B12 by *E. coli* requires a functional *TonB* product. By using a temperature-sensitive *tonB* mutant, it can be demonstrated that as *tonB* function declines on shifting to the non-permissive temperature, the cell rapidly loses sensitivity to colicins B, D and I and the rate of B12 translocation rapidly falls. Cloning of the *tonB* gene has revealed that it codes for a 36,000 molecular weight polypeptide. The porin proteins are thought capable of a gating response on application of a voltage (p. 88). It is perhaps not too fanciful to propose that the TonB product may have a role in the establishment of such electric fields.

Analysis of the outer membrane protein profile of clinical isolates of *E. coli*, rather than the more usual laboratory strains has shown that although in general the polypeptide pattern observed is similar, significant differences occur. Encapsulated strains of clinical isolates possess an additional 40,000 molecular weight polypeptide in the outer membrane, which

Table 3.4: Outer Membrane Proteins of *E. coli* Encoded by Exogenous DNA

Protein	Relative molecular mass	Coding DNA	Function
TraT	25,000	F-factor	Surface exclusion; prevents the formation of stable mating aggregates with other competent strains
Protein 2	38,000	Prophage PA2	Replacement of the PA-2 binding protein (OmpC) with a non-phage binding equivalent porin to prevent superinfection
Iss	(33,000?)	ColV-K94	Increased resistance to serum complement
'Resistance porins'	(37,000)	R-factors	Replacement of existing porin

because of its association with capsule formation is designated protein K. It is closely related to the major porin proteins, in terms of its amino acid composition and *N*-terminal analysis.

Changes in the outer membrane protein profile can also be caused by the introduction of additional DNA (Table 3.4). The incorporation of the lysogenic phage PA2 into the host DNA of *E. coli* K12 results in the replacement of the OmpC protein (the PA2 receptor) and the OmpF porin, by a novel porin encoded on the phage genome, designated protein 2. This new polypeptide (M_r 38,000) is incapable of binding PA2 phage and is presumably a mechanism to prevent superinfection. The presence in the host bacteria of a factor promoting conjugation, e.g. the F-factor, results in the synthesis of the surface exclusion proteins specified by the *traT* and *traS* genes. The TraT polypeptide (M_r 25,000) is located at the surface of the outer membrane and serves to prevent the formation of stable mating pairs with other donor cells. An additional property of this protein is that it appears to increase the resistance of the bacterium to the complement-directed bactericidal action of serum (p. 157). The acquisition of the ColV-K94 plasmid also increases the serum-resistance of *E. coli* strains. Cloning techniques have been used to show that the resistance is due to the *iss* gene product, which presumably acts in a manner similar to *traT*. Strains of *E. coli* K12 harbouring the ColV K94 plasmid contain a new major outer membrane protein (M_r 33,000). Drug resistance factors of the *N* incompatibility group induce the replacement of the existing pore protein of *E. coli* B with a series of alternative porins. It has been suggested that this change is sufficient to account for the increased drug resistance of organisms bearing such plasmids.

Only limited enzymic activity is associated with the outer membrane. A major outer membrane protein of enteric bacteria, protein *a*, is endowed with some proteolytic activity (p. 89). The outer membrane of *E. coli* contains detectable levels of phospholipase A, a 28,000 polypeptide requiring divalent ions for activity. In *Neisseria meningitidis*, tetramethyl-phenylenediamine oxidase activity, the basis of the well known 'oxidase reaction' is present in the outer membrane.

Periplasmic Proteins of Gram-negative Bacteria

The region in the Gram-negative envelope situated between the cytoplasmic and outer membrane is known as the periplasmic space (Figure 3.2). The space constitutes some 20–40 per cent of the total cell volume. The periplasm is distinguished by a distinct ionic composition and the possession of a unique series of proteins. The restrictive permeability properties of the outer membrane ensure that periplasmic components cannot easily leak into the environment. Certain of the proteins have a catalytic function, converting a variety of compounds into a form amenable to translocation by specific carriers present in the cytoplasmic membrane. A second group of periplasmic enzymes catalyse the destruction of certain antibacterial agents able to penetrate the outer membrane. The periplasm also contains a series of nutrient-binding proteins which are essential components of certain active-transport systems. The unique ionic milieu compartmentalised within the periplasm creates a demonstrable Donnan potential which may, in conjunction with TonB, play a role in translocation and chemotactic signalling events in the outer membrane.

The periplasmic contents are released by a mild cold osmotic shock procedure. In essence this consists of plasmolysing the bacteria by hypertonic sucrose/EDTA, followed by resuspension in cold distilled water. This releases approximately 4 per cent of the total cell protein; the bacteria remain viable, though unable to perform any function dependent upon the lost periplasmic binding protein.

Scavenging Hydrolytic Periplasmic Enzymes. These proteins are generally involved in the degradation of metabolisable compounds which are too large or too highly charged to pass through the cytoplasmic membrane. Table 3.5 contains a representative, though by no means complete, list of the enzymes present. Many of the enzymes are membrane associated, presumably present at the exterior face of the cytoplasmic membrane. The substrates of the periplasmic enzymes cannot normally be taken up by the cell, but after enzymic degradation the reaction products can be translocated to the cytoplasmic interior.

Detoxifying Enzymes. Exposure to antimicrobial agents will rapidly select drug-resistant mutants. The commonest mechanism of drug resistance is

Table 3.5: Periplasmic Enzymes of Gram-negative Bacteria

Enzyme	M, of substrate (approx)	Organism
Scavenging proteins		
Asparaginase	132	
Alkaline Phosphatase	260	
Acid hexose phosphatase	260	
Non-specific acid phosphatase	260+	
5'-nucleotidase		*E. coli*
(UDP-glucose hydrolase)	350	
3'-nucleotidase		
(cyclic phosphodiesterase)	350	
ADP glucose hydrolase	590	
Enzymes catabolising deoxyribonucleosides		
Deoxyribomutase	214	
Deoxyriboaldolase	214	
Purine deoxyribonucleoside	276	*E. coli*
phosphorylase		
Deoxythymidine phosphorylase	242	
Nucleases		
Ribonuclease 1	1200+	*E. coli, S. typhimurium*
Deoxynuclease 1	1200+	
Detoxifying enzymes		
β-lactamase	380	*E. coli, S. typhimurium*
Aminoglycoside-3'-phosphotransferase		*E. coli, Klebsiella* sp.
Aminoglycoside-2'-acetylase	581	*E. coli, Proteus* sp.
Aminoglycoside-2'-adenylase		*E. coli, Pseudomonas aeruginosa*
Alkyl sulphohydrolase	166+	*Pseudomonas*

the production of enzymes capable of degrading or modifying the anti-microbial agent, in such a way as to render the agent harmless. The production of the β-lactamase enzyme which degrades penicillins and cephalosporins (p. 21) is common in both Gram-negative and Gram-positive bacteria. Gram-positive bacteria produce the enzyme directly into the environment, but in many Gram-negative bacteria much of the enzyme remains in the periplasm (Table 3.5). Similarly enzymes capable of inactivating aminoglycoside antibiotics (streptomycin, kanamycin or gentamicin) by adenylation, phosphorylation or acetylation are also present in the periplasm. Within the periplasm of certain pseudomonads are found enzymes capable of degrading toxic detergents. The confining of detoxifying enzymes in the periplasmic space means that the inactivating enzyme concentration is greatest where it is needed most, close to the cell. Only

Table 3.6: Specific Binding Proteins of Gram-negative Bacteria

Ligand	M_r of ligand	M_r of binding protein	Dissociation constant (µM)	Organism
Anions				
Sulphate	98	31,000	20	*S. typhimurium*
Phosphate	98	41,000	0.08	*E. coli*
Amino acids				
Leucine ⎫	131		0.6 ⎫	
Isoleucine ⎬	131	36,000	0.6 ⎬	*E. coli*
Valine ⎭	117		10 ⎭	
Glutamine	146	25,000	0.3	*E. coli*
Lysine ⎫	146		3 ⎫	
Arginine ⎬	174	27,000	1.5 ⎬	*E. coli*
Ornithine ⎭	132		5 ⎭	
Histidine	155	26,000	1	*S. typhimurium*
Sugars				
Arabinose	150	38,000	2	*E. coli*
Ribose	150	30,000	0.2	*E. coli*
Maltose	342	37,000	160	*E. coli*
Galactose ⎫	180	35,000	1 ⎫	⎧ *E. coli*
Glucose ⎭	180		0.5 ⎭	⎨ *S. enteritidis*
Ribose	150	31,000	0.3	*S. typhimurium*
Vitamins				
Thiamin	301	40,000	0.05	*E. coli*
B12	1,355	22,000	0.006	*E. coli*

drug molecules that are directly threatening the life of the organism are degraded. This confers a great advantage to Gram-negative bacteria since, unlike Gram-positive bacteria, they do not need to produce large quantities of drug-destroying enzyme which is constantly being lost to the environment.

Periplasmic Binding Proteins. Closely associated with the periplasmic face of the cytoplasmic membrane is found a series of soluble proteins (M_r 26,000–43,000), which although devoid of enzymic activity are capable of binding specific solutes (K_d 10^{-8} to 10^{-5}). The first binding protein of this type to be isolated was the sulphate-binding protein of *S. typhimurium* (Table 3.6). The sulphate-binding protein is a single polypeptide and is present in the bacterium in large quantities, representing almost 1 per cent of the total cell protein (approximately 2×10^4 molecules/cell). The protein bears a single sulphate binding site with a K_d of 2×10^{-5}. The osmotic shock procedure releases 80 per cent of the sulphate binding protein and this inhibits sulphate uptake by 80 per cent. The K_m for sulphate uptake (2×10^{-5}M) is the same as the K_d of the sulphate binding protein. Mutation of the *cysA* and *cysB* gene reduces both the level of binding protein and the rate of sulphate uptake. These observations provide strong, though indirect, evidence that the binding protein is an essential component of the sulphate transport system of *S. typhimurium*. The phosphate binding protein of *E. coli* is also a single polypeptide chain containing a single ion binding site (K_d 0.8×10^{-7}). The protein functions as part of a high affinity phosphate transport system.

A wide variety of amino acids and sugars also possess specific periplasmic binding proteins (Table 3.6). In the main they are single polypeptides each bearing a single high affinity binding site. These proteins serve as the first step in specific transport pathways, which permit the accumulation by the cell of that particular ligand. Direct interaction of the periplasmic binding proteins with outer membrane porins may also be important in solute uptake. Permeation of maltodextrins through the bacterial envelope has been shown to involve the co-operation of the periplasmic MalE binding protein with the LamB porin. A binding protein for thiamin can be purified from *E. coli.* The protein has a single high affinity thiamin binding site (K_d 5×10^{-8}) and is thought to be a component of the thiamin transport system. The periplasm of *E. coli* also contains a binding protein for vitamin B12. The protein is readily released by osmotic shock and can be readily distinguished from the BtuB protein.

Essential solutes able to permeate the outer membrane are rapidly sequestered by the appropriate periplasmic binding protein. This lowers the effective free solute concentration facilitating further solute diffusion. The binding protein is capable of mediating the rapid transfer of the solute to the appropriate cytoplasmic membrane permease which transports the ligand to the cell interior.

Biosynthesis and Assembly of Periplasmic and Outer Membrane Proteins

A number of features distinguish the biosynthesis of outer membrane and periplasmic proteins from the cytoplasmic proteins. Examination of the sensitivity of cytoplasmic and outer membrane polypeptide synthesis to inhibitors of ribosome function suggests that the outer membrane proteins utilise different biosynthetic machinery. The biosynthesis of envelope proteins is strikingly more resistant to puromycin and kasugomycin than the transcription of cytoplasmic protein. In contrast, the synthesis of cytoplasmic proteins continues at levels of tetracycline and sparsomycin that halt the biosynthesis of envelope proteins. Differential sensitivity to the action of the drugs can also be observed among individual envelope proteins; OmpA biosynthesis is more resistant to kasugomycin, sparsomycin and chloramphenicol than any of the other outer membrane proteins. In contrast lipoprotein synthesis is chloramphenicol-sensitive but very resistant to puromycin. At high levels of puromycin (up to 500µg/ml) lipoprotein biosynthesis continues, although all other envelope protein biosynthesis is halted. Inhibitors of *m*RNA synthesis, e.g. rifampicin, effectively halt cytoplasmic protein synthesis, with little effect on envelope protein biosynthesis. Since rifampicin blocks the initiation of *m*RNA synthesis the stability of *m*RNAs can be estimated by measuring the rate of protein synthesis after addition of the drug. It appears that the *m*RNAs coding for envelope proteins are, on average, 2.5 times more stable than those of cytoplasmic proteins.

Outer membrane and periplasmic proteins are preferentially synthesised on membrane-bound polyribosomes, suggesting that the nascent polypeptide is translocated across the membrane during protein synthesis. Treatment of *E. coli* cells with low levels of toluene breaks down the envelope permeability barrier. Toluene-treated cells are permeable to nucleoside phosphates. It is possible to establish a toluenised *E. coli* system which synthesises only membrane proteins; synthesis being entirely dependent upon the addition of exogenous ATP. When the outer membrane lipoprotein is synthesised in this system, the product appears to be larger than normal. Determination of the molecular weight and amino acid sequence of this prolipoprotein reveals the presence of an additional 20 amino acid sequence at the *N*-terminal. Isolated lipoprotein-specific *m*RNA can be used in cell-free protein synthesis; examination of the product showed it to be identical to the prolipoprotein produced by toluenised cells. Lipoprotein is not the only outer membrane protein to possess such an additional sequence; in permeabilised *E. coli* cell preparations OmpA, OmpC, OmpF and TraT are synthesised in precursor forms, containing about 20 extra amino acid residues. *In vitro* synthesis of the LamB protein and several periplasmic proteins (the binding proteins for maltose, arabinose, ribose, leucine, leucine/isoleucine/valine, histidine, lysine/arginine/ornithine and histidine) also results in the formation of a larger precursor protein. The presence of an additional *N*-terminal sequence has also been detected in precursor forms of certain periplasmic enzymes, e.g. alkaline phosphatase and TEM β-lactamase. However it would be wrong to assume that all periplasmic and envelope proteins contain *N*-terminal extensions.

These observations are strong evidence in favour of the 'signal' hypothesis of Blobel and Dobberstein, who proposed that secreted proteins are produced with an extra *N*-terminal sequence that facilitates the passage of the nascent polypeptide through the cytoplasmic membrane. The *N*-term-

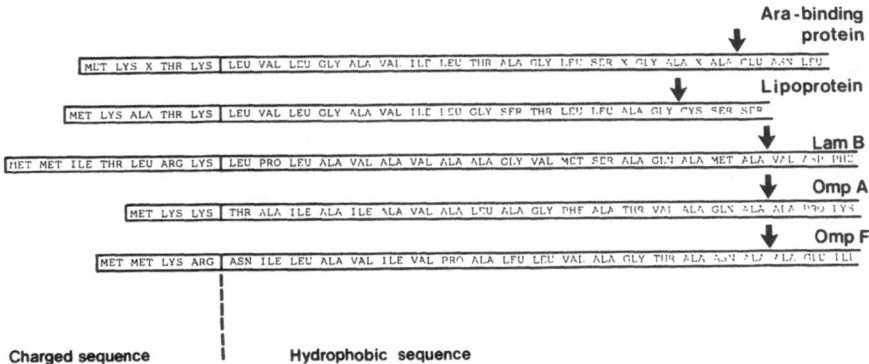

Figure 3.30: *N*-terminal Signal Sequence Extensions of the Arabinose-binding Periplasmic Protein, and the Lipoprotein, LamB, OmpA and OmpF Outer Membrane Proteins of *E. coli*. The arrows denote the site of proteolytic cleavage

inal extension being split off later, at the outer face of the cytoplasmic membrane.

Amino acid and DNA sequencing techniques have been used to determine the primary structure of many bacterial signal sequences. Comparison of the sequence reveals little homology, with the notable exception of lipoprotein and the arabinose binding protein, whose signal peptides are very similar (Figure 3.30). However, all the signal peptides contain a positively charged *N*-terminal region of the two to eight amino acids, followed by a long sequence of hydrophobic amino acid residues. The amino acid situated at the cleavage site bears a short side chain. Biophysical studies indicate that the peptide exists in a highly ordered helix and that the hydrophobic sequence is sufficient to span the cytoplasmic membrane. Genetic studies have shown that a specific sequence is essential for the export of many proteins. Mutations that introduce a charged amino acid into the hydrophilic region of the signal sequence of LamB, maltose binding protein or β-lactamase can cause the precursor form to accumulate in the cytoplasm.

It is thought that the positively charged section of the signal polypeptide interacts with the negatively charged inner surface of the cytoplasmic membrane, effectively fixing the polyribosome to the membrane. The hydrophobic section then progressively inserts into the membrane facilitated by hydrophobic interaction with the lipid bilayer. The cleavage site of the signal sequence eventually becomes exposed at the periplasmic face of the membrane. Removal of the signal peptide is a relatively late event, occurring after the polypeptide reaches at least 80 per cent of its final length. One periplasmic enzyme, a β-lactamase, appears in the periplasm as the precursor, which is only then modified by proteolysis.

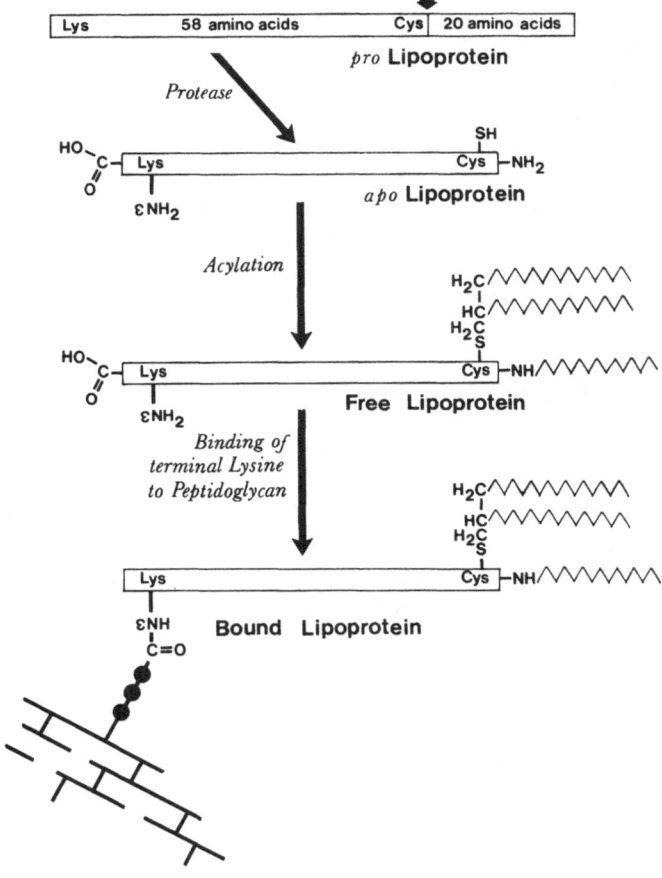

Figure 3.31: Post-translational Modification of the Pro-lipoprotein in *E. coli*. An extracellular protease cleaves off the 20 amino acid signal sequence necessary for translocation across the cytoplasmic membrane, revealing the *N*-terminal cysteine. This is modified and acylated to give the free-lipoprotein, which may subsequently bind to the peptidoglycan

After translation of lipoprotein mRNA the product must undergo an ordered series of post-translational modification before the final product inserts into the outer membrane (Figure 3.31). After cleavage of the signal polypeptide, the cysteine residue at the new *N*-terminal is extensively modified. Firstly the free amino group is acylated forming an amide-linked fatty acid, usually palmitic acid, but palmitoleic and *cis*-vaccenic acid may also be detected. A diglyceride is then linked to the sulphydryl group of the *N*-terminal cysteine. The ester-linked fatty acids differ little from those of general cellular phospholipids suggesting that the diglyceride derives from general phospholipid biosynthesis. Some of the lipoprotein molecules are then covalently linked to the peptidoglycan. The ε-amino group of the lipoprotein C-terminal lysine forms a peptide bond with the carboxyl group of diaminopimelic acid in the peptidoglycan. The conversion into the bound form of lipoprotein is apparently not essential for the insertion into the outer membrane, since the free form is first inserted into the outer membrane before linking to the peptidoglycan. It appears that the acylation of the apolipoprotein (Figure 3.31) is an essential requirement for incorporation of the molecule into the outer membrane, the fatty acids providing the necessary degree of hydrophobicity.

Two important unanswered questions are how the other outer membrane proteins transfer from the cytoplasmic membrane to the point of insertion into the outer membrane and how the synthesis of outer membrane proteins is regulated. It is unclear as to whether the precursors are processed to release the completed protein into the periplasm to find its own way to the outer membrane or whether additional, yet to be identified, translocatory mechanisms are involved. The kinetics of insertion of newly synthesised proteins can be determined by using pulse-chase techniques. Although proteins are rapidly integrated into cytoplasmic membrane (about 1 min) the insertion of outer membrane proteins is much slower (about 3 min). The release of proteins into the periplasm requires about 10 min, which suggests that outer membrane proteins do not reach their final destination via the periplasm. The synthesis of the OmpF porin of *S. typhimurium* is repressed by sodium chloride. If a sodium chloride-repressed culture is resuspended in salt-free medium then the sites of porin-insertion can be visualised under the electron microscope using ferritin-labelled anti-OmpF antibody. Newly synthesised OmpF protein appeared at numerous but discrete sites, resembling those at which LPS is inserted. The most plausible explanation of the specific insertion of polypeptides into the outer membrane is the possession by these proteins of a high affinity towards other outer membrane components, probably LPS. The porin proteins and OmpA exhibit a great affinity for LPS; binding between OmpC/OmpF and LPS resists all but the most vigorous extraction procedures. A specific high-affinity interaction between the N-terminal sequence of the OmpA protein and LPS has been reported. Deep rough mutants (R_d, R_e; Figure 3.16) of both *S. typhimurium* and *E. coli* show marked reduction in the levels of major outer membrane proteins. It is unclear as to why loss of much of the LPS core region should result in such a drastic modification of the outer membrane profile, unless their biosynthesis or translocation is related. Newly synthesised LPS is rapidly translocated to the exterior face of the outer membrane. The presence of pore-type proteins in the area of the cytoplasmic membrane could pose severe problems in maintaining integrity of the bilayer. An attractive hypothesis is that the high affinity of

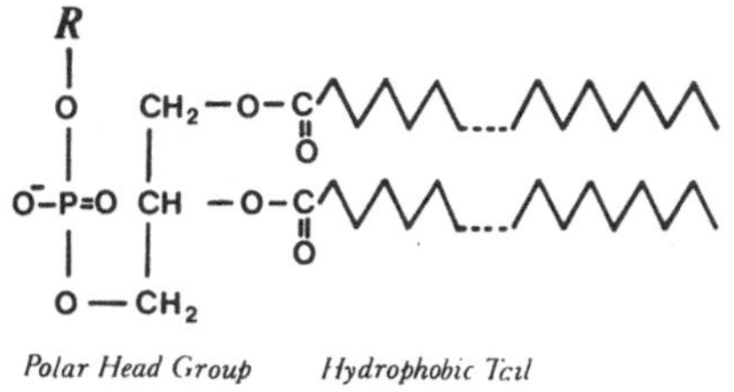

Figure 3.32: Structure of Bacterial Phospholipids

the LPS for the protein ensures a rapid unidirectional cotranslocation of the protein:LPS complex. Although deep rough LPS is readily translocated, it may be less able to form a suitable LPS:protein complex, the subsequent accumulation of polypeptides in the cytoplasmic membrane repressing the synthesis of new outer membrane protein. If the assembly of the outer membrane proteins is coupled with the insertion of newly synthesised LPS, then it is probable that both components utilise the regions of adhesion between inner and outer membrane.

Outer Membrane Phospholipids

The predominant phospholipid of *E. coli* and *S. typhimurium* is phosphatidylethanolamine (PE) with some phosphatidylglycerol (PG) and diphosphatidylglycerol (diPG), sometimes known as cardiolipin (Figure 3.32). The phospholipid composition of cytoplasmic and outer membranes is relatively similar: about 60 per cent of the total phospholipid being present in the outer membrane. The outer membrane of *S. typhimurium* is greatly enriched for PE and unlike the cytoplasmic membrane contains negligible diPG. The phospholipid composition is not dramatically modified by changes in growth conditions. The ratio PE to PG + diPG appears constant. In stationary phase cultures the level of diPG tends to increase at the expense of PG. The acyl substituents of bacterial phospholipids are essentially the same consisting primarily of palmitic, myristic, palmitoleic and *cis*-vaccenic acids (Table 3.7). The outer membrane phospholipids, predominently PE, have a higher ratio of saturated to unsaturated fatty acids. The increased proportion of unsaturated fatty acids in the cytoplasmic membrane is probably due to its greater content of PG enriched in *cis*-vaccenic acid. When *E. coli* is grown at 37°C, the majority of fatty acids are C_{16}, but lower temperatures increase the proportion of unsaturated acids. Conversely higher temperatures increase the degree of saturation of the acyl chains. Cylopropane derivatives of palmitoleic and *cis*-vaccenic acids (Table 3.7) are more abundant in stationary-phase cultures.

Table 3.7: Fatty Acids of Enteric Bacteria

Saturated

$$CH_3(CH_2)_n — COOH$$

n = 10, lauric; n = 12, myristic;

n = 14, palmitic; n = 16, stearic

Unsaturated

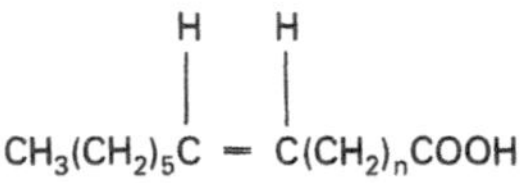

$$CH_3(CH_2)_5\overset{\overset{\displaystyle H}{|}}{C} = \overset{\overset{\displaystyle H}{|}}{C}(CH_2)_nCOOH$$

n = 7, palmitoleic acid;

n = 9, *cis*-vaccenic acid

Cyclopropane

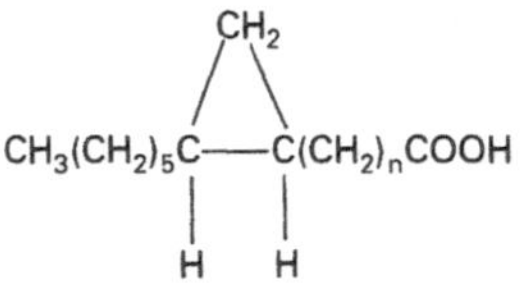

$$CH_3(CH_2)_5\overset{}{C}—\overset{}{C}(CH_2)_nCOOH$$ with CH₂ bridge above and H below each carbon

n = 9, lactobacillic acid

A number of membrane enzymes are functionally dependent on the presence of phospholipid, including the membrane-bound ATPase, NADH oxidase, C_{55} isoprenoid-alcohol kinase, diglyceride kinase and *sn*-glycerol 3-phosphate acetyltransferase. The UDP-galactosyltransferase of LPS biosynthesis is active as a ternary complex with PE and LPS. The phospholipid head group is also involved in certain transport phenomena. PG is able to activate the enzymes II of the phosphotransferase system responsible for the uptake of certain sugars by the bacterium. Mutation of the *pss* gene produces organisms defective in phospholipid biosynthesis. Such mutants are hypersensitive to antibiotics, in particular aminoglycosides. This suggests that the phospholipids are an essential component of the outer membrane permeability barrier. The ratio of dipolar ionic to negative polar head groups in the membrane is vital to normal function. In wild type strains negatively charged phospholipids comprise about 20 per cent of the total. Mutants containing 40 per cent negative phospholipid grow more slowly and if PE levels fall below 50 per cent growth ceases. In contrast mutants unable to synthesise diPG grow normally since such mutations have no effect on charge distribution.

The outer membrane must contain some fluid and some non-fluid fatty acids at all temperatures, leading to a membrane with a mix of saturated and unsaturated acids. Changes in the temperature of the environment are rapidly reflected in altered fatty acid composition. At lower temperatures, the minimum amount of saturated fatty acid required is less than that needed at higher temperatures. A homeostatic mechanism ensures that the composition of the membrane is adjusted to produce a structure with the fluidity characteristic best suited for optimum membrane function. The incorporation of cyclopropane or branched fatty acids will achieve the same end result as unsaturated fatty acids. The movement of lipids within the plane of the membrane is quite rapid. The coefficient of lateral diffusion can be determined by using spin-labelled fatty acids and found to be 3.2×10^{-8}cm/s in *E. coli*, sufficiently rapid to permit a single phospholipid to move from one end of the bacterium to the other in about two seconds (3.5 μm/s). It has been suggested that the PE molecules of the outer membrane are predominantly located in the inner leaflet of the outer membrane, forming a monolayer with the polar groups of PE facing the cell interior (Figure 3.2), while the outer leaflet contains mainly protein and LPS. A true lipid bilayer will therefore only exist in small patches.

Phospholipid Biosynthesis

Phospholipids are assembled from three components, fatty acids, *sn*-glycerol 3-phosphate and serine. Fatty acid biosynthesis in *E. coli* is carried out by a series of soluble cytoplasmic enzymes (Figure 3.33). The basic building block of fatty acid biosynthesis is acetyl-coenzyme A (acetyl-CoA). Malonyl-coenzyme A is the first intermediate in fatty acid biosynthesis, in the presence of soluble enzyme fractions being rapidly converted into fatty acid. This has an absolute requirement for CoA. Acetyl- and malonyl-CoA themselves do not participate directly in the donation of acetyl and malonyl residues. The acyl intermediates are instead bound to a small polypeptide (M_r 9,600), the acyl carrier protein (ACP). Acyl residues bind to the sulphydryl group of a 4-phosphopantothine group attached to serine at position 36. ACP performs a function unique amongst known proteins, acting as a carrier for all the acyl groups necessary for the biosythesis of fatty acids. The malonyl- and acetyltransferases only transfer the respective substrates to ACP. β-ketoacetyl ACP synthetase is specific for the thioester of ACP, but relatively non-specific for the chain length of the condensation partner of malonyl-CoA. The β-hydroxy-acyl ACP dehydrase is specific for ACP derivatives and the D-isomer of the substrate. The β-hydroxy-butyryl ACP dehydrase is specific for short chain length substrates (activity markedly declining beyond C_4 and non-existent at C_{10}), suggesting another enzyme is responsible for dehydration of longer chain acids. In the presence of excess malonyl-CoA, a pulse of acetyl-CoA is only incorporated into C-15 and C-16 of palmitic acid. This suggests that the initial condensation of acetyl-CoA and malonyl-CoA is followed by successive chain elongation through condensation of the developing fatty acid with an additional molecule of malonyl-CoA. Hence the product of the enol-ACP hydrolase reaction reacts repeatedly with malonyl-ACP until

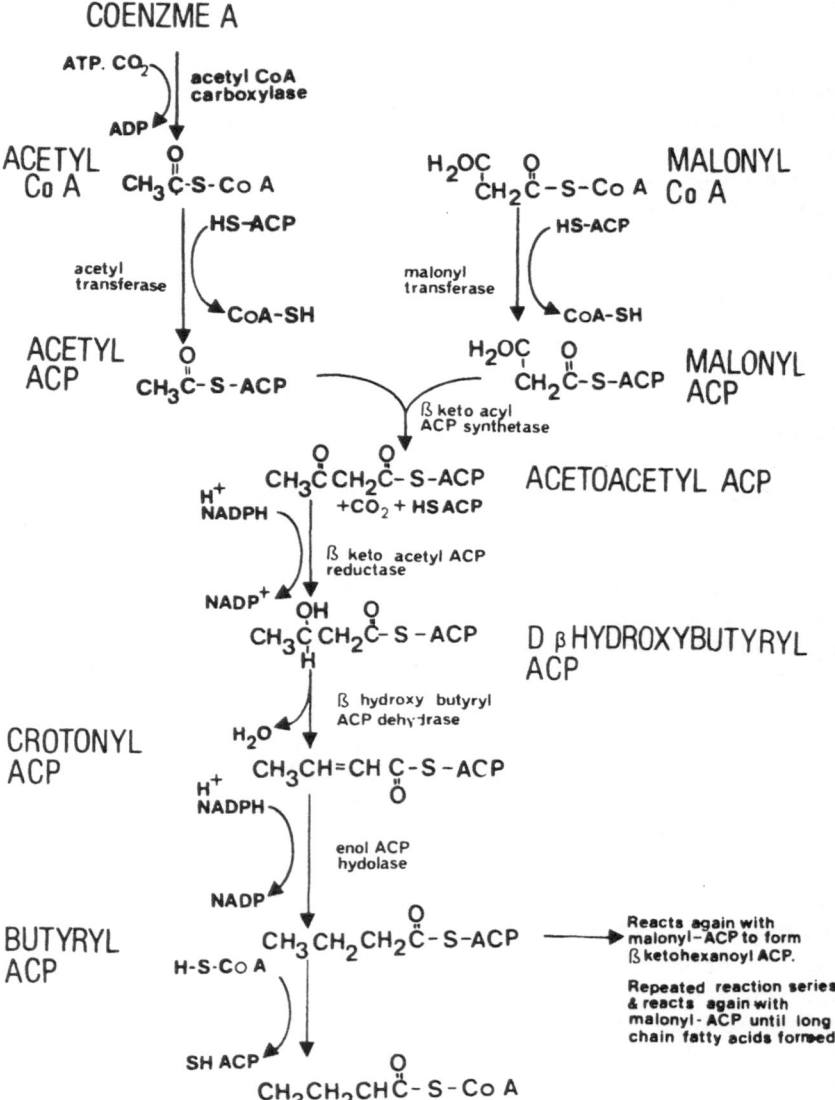

Figure 3.33: Biosynthesis of Fatty Acids in *E. coli*

the reaching of the desired chain length triggers the termination reaction. The terminal reaction involves transacylation with CoA, releasing the ACP for another round of synthesis. Propionyl-CoA will replace acetyl-CoA in the initial condensation with malonyl-CoA. This may account for the synthesis of odd number carbon fatty acids. Similarly branched-chain fatty acids are synthesised by a modification of the pathway, i.e. condensation of the appropriate short-chain fatty acyl-CoA with malonyl-CoA.

The second major phospholipid precursor is *sn*-glycerol 3-phosphate. In glucose-grown *E. coli* it is normally derived from dihydroxyacetone phosphate by a specific dehydrogenase utilising NADH or NADPH; *gps*A mutants unable to synthesise this enzyme cannot synthesise phospholipid unless provided with endogenous *sn*-glycerol 3-phosphate. An alternative route for the synthesis of *sn*-glycerol 3-phosphate is from glycerol and ATP, but in glucose-grown cells this is not an effective route, since the glycerol kinase responsible is inhibited by fructose 1,6-bisphosphate. Glycerol kinase mutants may be isolated that are insensitive to fructose 1,6-bisphosphate inhibition and these are able to synthesise *sn*-glycerol 3-phosphate from glycerol in the presence of glucose. A third precursor, L-serine is essential for the generation of the polar head group of PE.

All the reactions of phospholipid biosynthesis occur at the cytoplasmic membrane. *sn*-glycerol 3-phosphate is first acylated in the 1-position with a saturated fatty acid (Figure 3.34). The reaction product does not accumulate but is rapidly acylated to form phosphatidic acid. The rate of turnover of all the intermediates of phospholipid biosynthesis is extremely rapid. The phosphate acid reacts with cytidine triphosphate (CTP) or deoxyCTP (dCTP). Evidence suggests that the half-life of these intermediates is of the order of seconds. The formation of CDP-diglyceride (and dCTP-diglyceride) is the rate-limiting step of the pathway. CDP-diglyceride can donate its phosphatidyl moiety to either the hydroxyl group of L-serine or to the hydroxyl group at the 1-position of *sn*-glycerol 3-phosphate. It is at this branch point in the biosynthetic route at which the ratio of PE to PG + diPG is regulated. Phosphatidylserine and phosphatidylglycerophosphate do not accumulate; phosphatidylserine is rapidly decarboxylated to give PE and phosphatidylglycerophosphate dephosphorylated yielding PG. PE is a stable end product, but much of the PG is converted into diPG by a reaction, independent of CDP-diglyceride, which requires no metabolic energy.

The mechanism by which phospholipid inserts into the outer membrane is unclear. Pulse-chase experiments indicate that newly synthesised PE is first located in the inner leaflet of the outer membrane and later rotates ('flip-flops') through the lipid bilayer to become part of the external lipid leaflet. Attempts to visualise discrete sites of PE insertion into the outer membrane have failed. This is not surprising since the lateral diffusion time

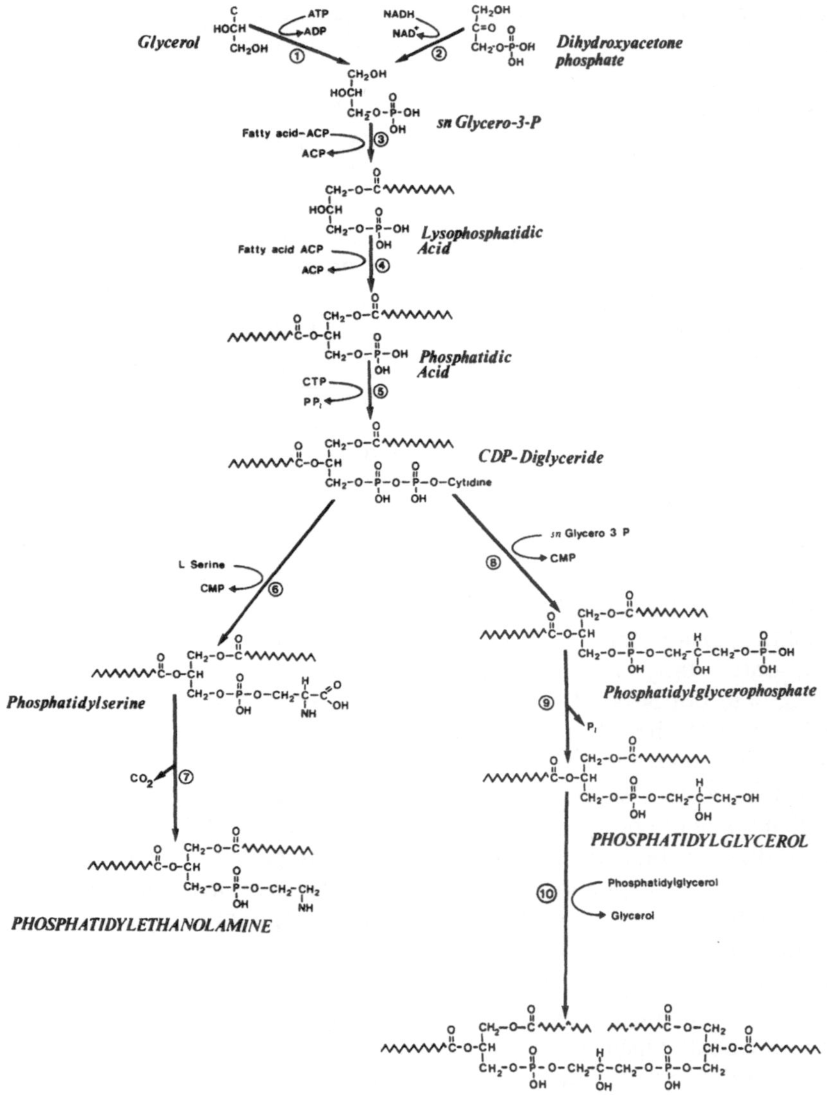

Figure 3.34: Biosynthetic Pathway of the Major Phospholipids in *E. coli*. Dihydroxyacetone phosphate, or less commonly glycerol are used in the synthesis of *sn*-glycerol phosphate (Steps 1 and 2). The *sn*-glycerophosphate is acylated to give phosphatidic acid (Steps 3 and 4), which reacts with CTP to form a liponucleotide derivative (Step 5). The CDP-diglyceride is situated at the branch point of the pathway. It can react with serine to produce phosphatidylserine (Step 6) and ultimately phosphatidylethanolamine (Step 7). Alternatively CDP-diglyceride can combine with *sn*-glycerol phosphate to yield phosphatidylglycerophosphate (Step 8) and then phosphatidylglycerol (Step 9). Phosphatidylglycerol can be further converted into diphosphatidylglycerol (Step 10) in a reaction that does not require CDP-diglyceride and, hence no metabolic energy

of lipids in the membrane is several orders of magnitude smaller than any practicable experimental system.

Membrane-derived Oligosaccharides

Within the periplasmic space of *E. coli* can be found a series of novel water-soluble oligosaccharides. The material is present in significant quantities, comprising 0.5 to 1 per cent of the dry weight of the cell. The oligosaccharides, of which at least three species have been detected, contain glucose as the only sugar (8–10 sugars, average M_r approx, 2,000). The glucose residue of one of the components is linked to succinic acid by an *O*-ester linkage and to *sn*-glycerol 1-phosphate and phosphoethanol-amine by a phosphodiester link to C-6 of the glucose residue. Another oligosaccharide of the series is devoid of succinate and ethanolamine. Study of the time course of radiochemical incorporation into the oligo-saccharide indicates that the *sn*-glycerol 1-phosphate moiety derives directly from the polar head group of polyglycerophosphatide (PG and diPG), hence the name membrane-derived oligosaccharides. The phospho-ethanolamine moiety is also thought to be derived from membrane phospholipids.

The PE component of bacterial membranes is relatively stable. In contrast, approximately a third, and in some strains all, of the PG pool turns over in one bacterial generation. The formation of the membrane-derived oligosaccharide and diPG synthesis accounts for three-quarters of the polyglycerophosphatide turnover *in vivo*. The necessity for this rapid turnover is unclear. The turnover of polyglycerophosphatides can be inhibited without any apparent deleterious effects upon the bacterium. Similarly mutants unable to synthesise membrane-derived oligosaccharides, show minimal polyglycerophosphatide turnover with no effect on cell growth. The membrane-derived oligosaccharide does not appear to have any direct role in fatty acid or phospholipid biosynthesis, but may have a transport function in phospholipid assembly that is not essential for growth in broth cultures.

Enterobacterial Common Antigen (ECA)

Enterobacterial common antigen is a common component shared by almost all wild-type strains of Enterobacteriaceae. It has been isolated from *Salmonella montevideo* and found to be a linear polymer of 1→4-linked *N*-acetyl-D-glucosamine and *N*-acetyl-D-mannosaminuronic acid, esterified to a limited extent by palmitic and acetic acids. These components account for about 70 per cent of the molecule, with an

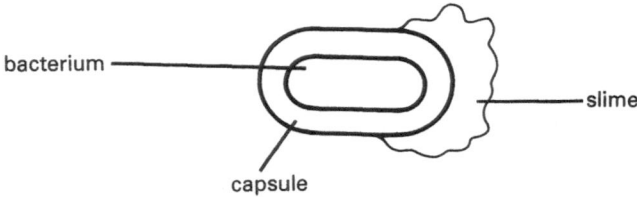

bacterium

slime

capsule

Figure 3.35: Representation of the Bacterial Capsule and Extracellular Slime Layers

uncharacterised lipid moiety making up the remainder. The molecule exists in two forms, a free ECA of M_r 2,700 present in most strains and an immunogenic form restricted to rough strains, in which the ECA is linked to LPS. Since it readily reacts with anti-ECA antibody it is thought that the antigen is located in the exterior leaflet of the outer membrane, particularly in those strains where it is directly linked to rough LPS stubs. The site of the ECA in the bacterial cell can be localised by using ferritin-labelled antibody. The ECA of rough strains of *E. coli* will specifically bind anti-ECA antibody raised in rabbits and the site of the ECA-rabbit antibody complex visualised under the electron microscope by subsequent treatment with ferritin-conjugated anti-rabbit IgG antibodies. The ferritin appears at the surface of the outer membrane. The anti-ECA antibody presumably reacts with the LPS-bound ECA. Since there is no apparent reason for the two forms of ECA, or the universality of ECA in enteric bacteria or any known function, ECA remains very much an enigma.

Exopolysaccharides of Gram-negative Bacteria

Many Gram-negative bacteria produce an extracellular polysaccharide either as discrete capsule (Figure 5.1) or as a layer of slime less obviously attached to the bacterial cell surface (Figure 3.35). Capsules are normally stable and can only be removed from the bacterial cells by vigorous shaking or by alkali extraction. As capsulated bacteria grown in liquid culture age there is a tendency for the capsule to dissolve, leading to a mixed population of capsulate bacteria (with and without slime), bacteria bearing slime; non-capsulate bacteria and free slime. Cultural conditions also affect the dimension of the observed polysaccharide layer. Depending on species and cultural conditions, the capsule may extend 0.2 to 1.0 μm beyond the outer membrane. Exopolysaccharides, are immunogenic and correspond to the K (Kapsel)-antigen.

Gram-negative exopolysaccharides are built from a large range of monosaccharides; neutral hexoses being the most common, in particular D-glucose, D-galactose and D-mannose followed by 6-deoxyhexoses, e.g. L-fucose and L-rhamnose. Pentose sugars are rarely present. Amino sugars are common, in particular N-acetyl-D-glucosamine and N-acetylgalactosamine, as are uronic acids, e.g. D-glucuronic acid. KDO, which was originally thought to be restricted to LPS can be identified as a component of exopolysaccharide in a limited number of E. coli strains. The sugars of the exopolysaccharides may contain phosphate. Many of the sugars contain ester-linked acyl groups, mainly O-acetyl groups, but also succinate or formate substituents.

Gram-negative Homopolysaccharide Capsules

Relatively few homopolysaccharide capsules have been studied in Gram-negative bacteria. Important exceptions are the polymers of N-acetyl-neuraminic acid, termed sialic acids. The capsules of E. coli K1 strains and *Neisseria meningitidis* B and C have been shown to be sialic acid polymers (Figure 5.7). These capsules are poorly immunogenic presumably because of the presence of sialic acid residues in mammalian tissues and this has important implications in the pathogenic potential of the organism (p. 158). Form variation of the E. coli K1 antigen can be detected; in the OAc$^+$-form polysaccharide, approximately 90 per cent of the neuraminic acid residues are O-acetylated, whereas in the OAc$^-$- variant there is little or no O-acetylation. *Alcaligenes faecalis* var *myxogenes* produces a succinylated glucan capsule, the glucose being β-linked in the 1→3, 1→4 and 1→6 configurations. The exopolysaccharide of *Agrobacterium tumefaciens* (and many other agrobacteria) consists of a β1→2 linked glucan. A similar glucan is present in *Rhizobium* species. *Acetobacter xylinum* produces extracellular fibres, consisting of about 600 glucose residues, which closely resembles cellulose.

Gram-negative Heteropolysaccharide Capsules

The bulk of the described Gram-negative exopolysaccharides fall into this category. The composition of many of the K-antigens of E. coli have been characterised and, with the exception of K1, shown to be heteropolysaccharides (Figure 3.36). Genetic analysis has revealed two distinct gene loci controlling exopolysaccharide synthesis in E. coli. Synthesis of the K-antigens of strains in the O groups O8, O9 and O20, e.g. K8, K9, K17 and K57, are closely linked to the *rfb his*-linked gene cluster (44' on the E. coli chromosome). The production of the remaining K antigens is linked to a separate *serA*-linked locus designated *kpsA* (61' on the E.coli chromosome). The presence of two distinct gene loci permits the construction of E. coli strains with both types of K-antigens on a single cell. Many of the E. coli K-antigens, e.g. K1, K2, K5, K12 and K13, have a relatively low molecular weight and a high charge density. Others, e.g. K29, are

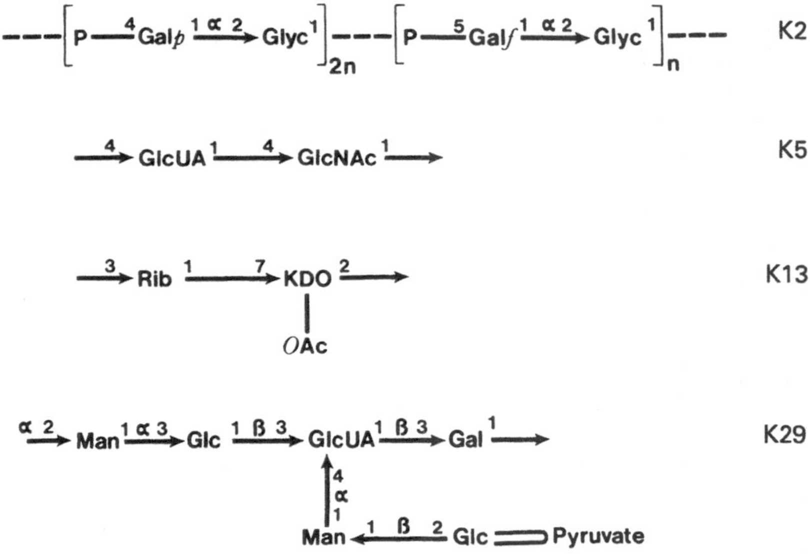

$$---\left[\text{P}\xrightarrow{4}\text{Gal}p\xrightarrow{1\ \alpha\ 2}\text{Glyc}^1\right]_{2n}---\left[\text{P}\xrightarrow{5}\text{Gal}f\xrightarrow{1\ \alpha\ 2}\text{Glyc}^1\right]_n--- \qquad \text{K2}$$

$$\xrightarrow{4}\text{GlcUA}^1\xrightarrow{4}\text{GlcNAc}^1\longrightarrow \qquad \text{K5}$$

$$\xrightarrow{3}\text{Rib}^1\xrightarrow{7}\text{KDO}^2\longrightarrow \qquad \text{K13}$$
$$\mid$$
$$O\text{Ac}$$

$$\xrightarrow{\alpha\ 2}\text{Man}\xrightarrow{1\ \alpha\ 3}\text{Glc}\xrightarrow{1\ \beta\ 3}\text{GlcUA}\xrightarrow{1\ \beta\ 3}\text{Gal}^1\longrightarrow \qquad \text{K29}$$
$$\uparrow 4$$
$$\alpha$$
$$1$$
$$\text{Man}\xleftarrow{1\ \beta\ 2}\text{Glc}\text{ ⊃ Pyruvate}$$

Figure 3.36: Structure of the Oligosaccharide Repeating Sub-units of Certain *E. coli* K-antigens. (*p* and *f* refer to the pyranose and furanose form of the sugars respectively.)

extremely heat stable, not being removed by boiling for 2 hours. In addition to the K-antigen, *E. coli* strains possess an extracellular slime layer. This is particularly evident in mucoid strains of *E. coli* K12, which produce copious amounts of slime. This material is termed M-antigen. The M-antigens are a series of acidic polysaccharide antigens, based on a hexasaccharide backbone of colanic acid (Figure 3.37). The terminal galactose of the side chain may be substituted by formaldehyde (methylidene substitution), acetaldehyde (ethylidene substitution) or pyruvate (carboxyethylidene substitution). Antigenically similar material to M-antigen has been detected in *Salmonella* and *Aerobacter* species.

The exopolysaccharide produced by *Xanthomonas campestris* has received much study. Xanthan gum, because of its unusual physical properties, has found considerable industrial usage. Xanthan gum dissolves readily in water producing highly viscous solutions at low concentrations. Aqueous solutions are highly pseudoplastic. Xanthan gums are used as gelling and viscofying agents, emulsifiers and stabilisers in foods, cosmetics, pharmaceuticals, textiles and printing. The exopolysaccharide is essentially a substituted cellulose polymer, the linear backbone of which consists of $\beta 1{\rightarrow}4$-linked glucose residues, and bears a trisaccharide side chain (Figure 3.38).

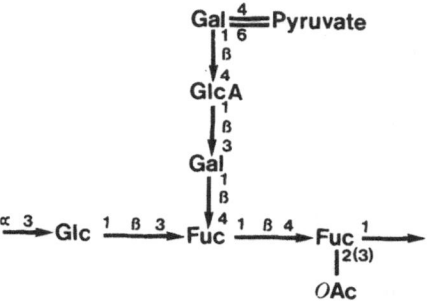

Figure 3.37: Structure of the Repeating Unit of Colanic Acid from *E. coli* K12

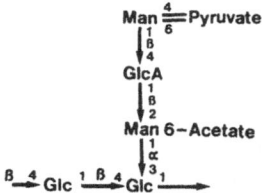

Figure 3.38: Structure of the Repeating Unit of the *Xanthomonas campestris* Exopolysaccharide (Xanthan Gum)

Alginate Exopolysaccharides of Gram-negative Bacteria

The exopolysaccharides produced by *Pseudomonas aeruginosa* and *Azotobacter vinelandii*, differ from those produced by other Gram-negative bacteria, in that the heterosaccharide appears to lack a repeating unit. These polymers consist of mannuronic acid and guluronic acid and closely resemble the alginates produced by marine algae.

When freshly isolated from soil *Azotobacter vinelandii* is highly mucoid, although this characteristic is rapidly lost on growth on laboratory medium. Analysis of the exopolysaccharide shows it to consist of mannuronic and galacturonic acid, with minor amounts of glucose, rhamnose, mannuronolactone and acetate. Approximately 20 per cent of the uronic acid groups bear *O*-acetyl groups (in contrast marine alginates are not acetylated). Most isolates of *Pseudomonas aeruginosa* do not produce large quantities of exopolysaccharides. However, strains isolated from pathological conditions, in particular respiratory tract infections that accompany cystic fibrosis, are almost invariably mucoid and may produce extraordinary large amounts of alginate. Alginate production is unstable, mucoid forms reverting to non-mucoid on subculturing. It is thought that the alginate provides some protection from host defences or antimicrobial agents. The polymer isolated from mucoid strains of *P. aeruginosa* is a partially acetylated alginate very similar to that of *Azotobacter*. The equivalent polymer produced by *P. stutzeri* is composed mainly of glucose and mannose and that of *P. fluorescens* entirely of mannose.

The alginates produced by *P. aeruginosa* and *A. vinelandii* are 1→4-linked unbranched co-polymers of β-mannuronic acid and its C-5 epimer α-L-guluronic acid. Partial acid hydrolysis suggests that the polysaccharide contains significant homopolymeric sequences of polymannuronic or polyguluronic acid, interspersed with other sequences that contain both monomers. The alginate is probably synthesised as a homopolymer of D-mannuronic acid and is subsequently modified by an extracellular epimerase converting some of the D-mannuronic acid residues into L-guluronic acid. *O*-acetylation of the alginate appears to be confined to the polymannuronic acid residues.

Biosynthesis of Exopolysaccharides

Many exopolysaccharide-producing bacteria can synthesise some extracellular polymer in the absence of utilisable carbohydrate. However, to produce optimum yield exogenous carbohydrate (glucose or fructose) must be available. The sugars are not directly incorporated into the polysaccharide, but must be converted into nucleoside diphosphate sugars. UDP-glucose plays a central role in the biosynthesis of several cell wall polymers, including exopolysaccharides. In addition to being the donor of glucosyl residues, UDP-glucose can be epimerised to UDP-galactose or oxidised to UDP-galacturonic acid, two important residues in exopolysaccharides. Bacteria actively synthesising exopolysaccharides possess high levels of UDP-glucose. GDP-mannose is of equal importance since not only is mannose a major component of exopolysaccharides, but GDP-mannose derivatives, GDP-mannuronic acid and GDP-fucose are also common. Present in the periplasm are sugar nucleotide hydrolases which may play a role in the processing or regulation of nucleotide sugar precursors. Precursors are also necessary for the acyl- and ketal-substituents common in completed exopolysaccharides.

Isoprenoid lipid carriers are used to convert the hydrophilic nucleotide diphosphate sugars into a form that is able to penetrate the cytoplasmic membrane. It will be recalled that isoprenoid sugar carriers are also used in the biosynthesis of peptidoglycan (p. 14) and LPS (p. 79). In actively growing bacteria there must be sufficient sugar-charged isoprenoid carrier to permit simultaneous synthesis of LPS, peptidoglycan and exopoly-saccharide, otherwise formation of one or more of the polymers would be limited. In some species exopolysaccharide is characteristically produced in late logarithmic phase or early stationary phase. This may be the result of insufficient isoprenoid carrier to supply concurrent polysaccharide synthesis by the three systems during logarithmic growth. It appears that peptidoglycan and LPS-biosynthesis have first call on available carrier.

Using cell-free preparations of *Klebsiella aerogenes* it can be shown that the oligosaccharide repeating unit of the exopolysaccharide is assembled in an analogous manner to that of similar envelope polymers (p. 10, p. 77) using identical transferase enzymes. Little is known concerning the final stage of exopolysaccharide biosynthesis. The mechanism(s) by which the oligosaccharide repeating units are transferred from the isoprenoid carrier to the developing polysaccharide and then extruded into the environment is not clear. The mode of attachment of exopolysaccharides to the underlying layers is also uncertain. Slime-forming (S1) mutants are frequently found in normally capsulated *Enterobacter aerogenes* strains, suggesting that a specific binding mechanism is present at the cell surface to attach the secreted polysaccharide. Presumably the S1 mutants lack this mechanism. The observation that the presence of a specific polypeptide in the outer membrane, designated protein K (p.95), correlates with capsule production in *E. coli*, may indicate a role for this protein in capsule attachment. High-resolution electron microscopy studies showing exopolysaccharide strands radiating from the bacterial surface have been cited as evidence for a distinct number of adherence sites, possibly the same sites implicated in LPS and protein export. Close examination of the electron micrographs reveals the presence of knob-like elements at the proximal ends of the capsular filaments. These structures are clearly not LPS, but whether or not they play a part in anchoring the exopolysaccharide is unclear.

Further Reading

Beacham, I.R. 'Periplasmic Enzymes in Gram-negative Bacteria',
   International Journal of Biochemistry (1979), *10*, 877–83
Berkeley, R.C.W., Gooday, G.W. and Ellwood, D.C. (eds) *Microbial
   Polysaccharides and Polysaccharases* (Academic Press, London, 1979)
Inouye, M. (ed) *Bacterial Outer Membranes* (John Wiley and Sons, New
   York, 1979)
Luderitz, O., Freudenberg, M.A., Galanos, C., Lehman V. Rietschel E.T.
   and Shaw, D.H. 'Lipopolysaccharides of Gram-negative bacteria',
   Current Topics in Membranesand Transport (1983), *17*, 79–151
Lugtenberg, B. and Van Alphen, L. 'Molecular Architecture and

Functioning of the Outer Membrane of *E. coli, Biochimica et Biophysica Acta* (1983), *737*, 51–115

Makela, P.H. and Mayer, H. 'Enterobacterial Common Antigen', *Bacterial Reviews* (1976), *40*, 591–632

Michaelis, S. and Beckwith, J. 'Mechanism of Incorporation of Cell Envelope Proteins in *E. coli'*, *Annual Reviews of Microbiology* (1982), *36*, 435–65

Neilands, J.B. 'Microbial Envelope Proteins Related to Iron', *Annual Reviews of Microbiology* (1982), *36*, 285–309

Orskov, I., Orskov, F., Jann, B. and Jann, K. 'Serology, Chemistry and Genetics of O and K Antigens of *E. coli'*, *Bacteriological Reviews* (1977), *41*, 667–710

Osborne, M.J. and Wu, H.P. 'Proteins of the Outer Membrane of Gram-negative Bacteria', *Annual Reviews of Microbiology* (1980), *34*, 369–422

Poindexter, J.S. 'The Caulobacters; Ubiquitous Unusual Bacteria', *Microbiological Reviews* (1981), *45*, 123–79

Rosen, B. *Bacterial Transport* (Marcel Dekker, New York, 1978)

Stoddart, R.W. *The Biosynthesis of Polysaccharides* (Croom Helm, London/Macmillan, New York, 1984)

Sutherland, I.W. *Surface Carbohydrates of the Prokaryote Cell* (Academic Press, London, 1977)

Sutherland, I.W. 'Biosynthesis of Microbial Exopolysaccharides', *Advances in Microbial Physiology* (1982), *23*, 79–149

Weckesser, J., Drews, D. and Mayer, H. 'Lipopolysaccharides of Photosynthetic Bacteria', *Annual Reviews of Microbiology* (1979), *33*, 215–39

4 SURFACE APPENDAGES: FLAGELLA AND FIMBRIAE

Bacterial Motility

The ability to perceive and respond to external stimuli give an organism a selective advantage over its competitors. Many bacterial species exhibit chemotactic behaviour, in which the organism moves towards or away from various chemicals or stimuli. Generally bacteria are attracted by nutrient compounds and repelled by potentially harmful agents. Bacterial chemotaxis has three distinct parts; receptors capable of detecting changes in the micro-environment, a system which modulates the chemotactic signal and finally a locomotory organelle.

Chemotactic Signal Reception

A variety of chemicals has been shown to elicit a chemotactic response. The observed response to any particular compound depends on the individual species, and in some cases will even differ between strains of the same species. Although some highly motile species are relatively insensitive to stimuli, most motile bacteria possess a battery of specific chemoreceptor systems, capable of detecting environmental changes. The triggering of the receptor initiates a series of reactions which ultimately direct the organism in the appropriate direction. Many simple nutrients, e.g. sugars and amino acids, have been shown to stimulate chemoreceptors. Each chemoreceptor is highly specific, only binding one specific chemical or its closely related analogue.

At least twelve chemoreceptor systems responsible for positive chemotaxis (i.e. movement towards increasing ligand concentration) to sugars are present in *E. coli* (Table 4.1). In *E. coli* and *Salmonella typhimurium* the chemotactic responses to galactose, maltose and ribose are sensitive to osmotic shock, a treatment known to release the periplasmic proteins (p. 97). This suggests that the specific sugar binding proteins present in the periplasmic space, which are known to be responsible for the initial stages of sugar transport, also have a role in the reception of the chemotactic signal. Mutation of the gene controlling synthesis of the galactose-binding protein results in an inability to respond to a galactose gradient, without affecting the response to other attractants or repellants. Mutants unable to transport galactose may nevertheless respond to the sugar and similarly other mutants may transport galactose without being attracted to it. Similar observations have been made with ribose and maltose. These observations indicate that the galactose-, maltose- and ribose-binding proteins serve as common receptors for both chemotaxis and transport, but

119

Table 4.1: Attractant Chemosensors in *E. coli.* Chemoreceptors are usually named after the compound they detect most efficiently. The 'serine' and 'aspartate' receptors are not characterised but may be periplasmic binding proteins and are probably not single receptors.

Sensor	Substrate	Receptor type	Transducing pathway
Glucose	Glucose	Enzyme II	
Mannose	Mannose	Enzyme II	
	Glucose	Enzyme II	
	2-deoxyglucose	Enzyme II	
Fructose	Fructose	Enzyme II	
N-acetylglucosamine	N-acetylglucosamine	Enzyme II	PEP pathway
Mannitol	Mannitol	Enzyme II	
Glucitol	Glucitol	Enzyme II	
Galactol	Galactol	Enzyme II	
Aryl-B-glucosides	Arbutin	Enzyme II	
	Salicin	Enzyme II	
Serine	Serine		
	Glycine	'Serine receptor'	Type I (*tsr*) MCPI
	Alanine		
	Threonine		
Aspartate	Aspartate		
	Glutamine	'Aspartate receptor'	Type II (*tar*) MCPII
	Methionine		
Maltose	Maltose	Periplasmic binding protein	Type II (*tar*) MCPII
Galactose	Glucose		
	Galactose		Type III (*trg*)
	Glycerol β galactoside	Periplasmic binding protein	MCPII or (I/II)
	Fucose		
Ribose	Ribose	Periplasmic binding protein	Type III (*trg*) MCPIII (or I/II)
Trehalose	Trehalose	—	—

subsequently the systems operate independently. The galactose, maltose and ribose chemoreceptors have been purified from *E. coli* and *S. typhimurium* and are closely related. The sugars exhibit a binding pattern consistent with one ligand molecule binding to each receptor.

The chemotactic receptors to glucose, fructose, mannose and certain other sugars are insensitive to osmotic shock (Table 4.1) and have been shown to be integral parts of the cytoplasmic membrane. These membrane-associated receptors have been shown to be identical with enzymes II, the substrate-specific components of phosphotransferase transport (the PEP system). The essential event during sugar translocation using the PEP system is the phosphoenolpyruvate-dependent phosphorylation of the

Table 4.2: Repellant Chemosensors in *E. coli*

Class	Receptor	Compound detected	Affected by *tsr* mutation	Transducer
1. Type I receptors	Indole	Indole	+	MCPI
		Skatole	+	
	Hydrophobic amino acids	L-leucine	+	MCPI
		L-isoleucine	+	
		L-valine	+	
		L-tryptophan	+	
		L-phenylalanine	+	
		L-histidine	—	
	Salicylate	Salicylate	+	MCPI ?
	Alcohols	Ethanol	+	MCPI ?
		iso-butanol		
2. Type 2 receptors				
Membrane ATPase	Metallic cations	Ni^{2+}	—	MCPII
		Co^{2+}	—	
3. $\Delta\bar{\mu}H^+$ receptors	Fatty acids	Acetate	+	MCPI
		Propionate	+	MCPI
		Butyrate	+	MCPI
	Benzoate	Benzoate	+	MCPI
	Hydroxyl	OH^+	+	MCPI
	Proton	H^+	—	?

solute. The transport step and phosphorylation requires two soluble enzymes (enzymes I and Hpr) and the enzyme II complex, containing enzyme IIA (which is sugar specific) and enzymes IIB (probably not sugar specific). Mutations in the enzymes II for glucose, mannose, fructose or galactose all exhibit defective chemotaxis to the respective compound. Mutants defective in enzyme I or Hpr exhibit poor chemotactic responses, i.e. the chemoreceptor and transport functions of the PEP system cannot be separated. The presence of an intact phosphorylation system for normal tactic responses may reflect a requirement of enzyme II for phosphorylated Hpr or direct phosphorylation to maintain high substrate affinity.

It is a reasonable assumption that transport systems arose before chemotaxis. The sharing of the binding proteins between an attractant chemosensor and the transport system for that ligand confers several benefits. Ability to move towards higher chemical concentrations is advantageous only if that compound can be efficiently transported and metabolised. The binding protein for a transport system would carry out the receptor requirement of an evolving chemosensor.

The magnesium-activated membrane-bound adenosine triphosphatase is important in the chemoreception of divalent ions, e.g. Ca^{2+}, Mg^{2+}, Zn^{2+} and Co^{2+} in *E. coli* and *S. typhimurium*. The cation receptor is part of the

BF_1 region of the enzyme. The discovery of a chemoreceptor function in BF_1 provides another example of the dual nature of the receptors involved in chemotaxis.

A large number of receptors for compounds eliciting repulsion chemotaxis are present in *E. coli* (Table 4.2) and similar patterns have been observed in *S. typhimurium.* It has proved difficult to identify specific receptors using direct binding assays since it appears that the affinity of repellant chemicals for their receptors is three or four times weaker than shown by attractant receptors for their ligands.

Several bacterial groups exhibit light-dependent alterations in their patterns of movement. Two types of photoresponses can be distinguished. In photokinesis, changes in light intensity cause alterations in the angular velocity of the organism. Photokinetic activity is normally maintained while the light intensity remains constant, i.e. adaptation does not occur. The action spectrum of photokinesis in photosynthetic organisms almost exactly matches that of the organism's photosynthetic pigments, suggesting that photokinesis is linked directly to the photocoupling reaction and lacks many of the properties of a true sensory process.

When exposed to light many motile bacteria show a phobic response, usually a sudden disturbance of the original behaviour pattern; either a stop response (a cessation of movement), a transient tumbling or a reversal of movement. These photophobic responses are rather varied, but each species exhibits a characteristic pattern dependent on the morphology of the organism and the action of its motor organelles. The photophobic response is determined by a temporal change in light intensity (i.e. up or down), rather than by the absolute magnitude of the stimulus and the response is all or none. An important feature of photophobic behaviour is that the organism readily adapts if the stimulus is sustained or frequently repeated. Photophobic responses are exhibited by photosynthetic purple bacteria, *Chromatia* and *Rhodospirillum,* and also in the halobacteria. While the latter respond to both increasing and decreasing light intensities the purple bacteria only respond to falling intensities. High intensity light pulses induce tumbling reactions in *S. typhimurium* and *E. coli* and backwards swimming in certain pseudomonads. Blue light is most efficient in eliciting a response and flavins have been suggested as the photoreceptor. As prolonged exposure to light leads to the loss of this response, photophobic taxis is not thought to be of great importance in the normal physiology of these organisms.

For both photosensor and photocoupling light quanta must be detected by specific receptor molecules. In blue-green algae and purple bacteria the same pigments, chlorophyll and bacteriochlorophyll respectively, serve as photoreceptors for photocoupling and phototaxis. Similarly bacteriorhodopsin, the photosynthetic pigment of *Halobacterium,* also mediates a step-down photophobic response. *Halobacterium* also possesses a second sensory photosystem, which mediates the step-up response independently of bacteriorhodopsin, which may involve carotenoid pigments.

Stimulus Transduction

Information detected by the chemoreceptors is transmitted to the loco-motory organelle through a heirarchical organisation of signalling components. Transducing components may receive imputs from several receptors and then pass the signal on, either to a secondary transducer or directly to the locomotory organelle. Although *trg* mutants of *E. coli* are defective for both ribose and galactose chemotaxis this locus does not map near any of the known genes for the ribose or galactose chemosensors. The *trg* product is believed to be a signalling element used by both receptors. Similarly the *tsr* gene product of *E. coli* has been shown to be required for the chemotactic response to a variety of compounds, including attraction to serine and repulsion by fatty acids, hydrophobic amino acids, indole, benzoate and low pH. Most *tsr* mutations destroy reaction to these compounds without affecting the response to other ligands. The *tsr* and *tar* transducers themselves bind the amino acids serine and aspartate respectively and therefore serve as the primary receptors for these compounds. Upon binding of the ligand the periplasmic receptor undergoes a conformational change and it is thought that this change initiates the reaction between the occupied chemoreceptor and the transducer. A similar system has been described in *Bacillus subtilis*, where a common signalling network services a series of amino acid receptors.

Methionine auxotrophs of *E. coli* exhibit normal chemotactic responses in the presence of the amino acid but if deprived of methionine they cease tumbling and do not react to attractants or repellants. Since this effect is observed only with methionine and occurs not only in enteric bacteria but also *B. subtilis*, an important role has been proposed for methionine in the processing of chemotactic signals. A derivative of methionine, *S*-adenosylmethionine, and not methionine itself, appears to mediate the 'methionine effect'. *S*-adenosylmethionine is a potent methylating agent and is thought to methylate a specific membrane protein, the methyl-accepting chemotaxis protein (MCP) in direct response to the chemotaxis signal. The methyl groups in the MCP appear to turn over with a time course equivalent to the rate of adaptation to the stimulus, suggesting that MCP is a component of the tumble generator or adaptation system.

E. coli mutants defective in the chemotactic response to all stimuli have been isolated. Lesions in eight genes (designated *che*) result in such behaviour (Table 4.3); falling into two broad phenotypic classes. *cheB* and *cheZ* mutants tumble incessantly. Mutation in *cheA*, *cheX*, *cheY* and *cheD* loci leads to reduced tumbling frequency. *cheC* mutants can exhibit either

Table 4.3: Function of Some of the Genes Involved in Chemotaxis of *E. coli*

Map position	Gene	Phenotype	Product mol. wt. $\times 10^{-3}$	
41.5'	*cheZ*	tumbles continually	24	mainly cytoplasmic
	cheY	fails to tumble	8–11	cytoplasmic
	cheB	tumbles continually	38	mainly cytoplasmic/methylesterase
	cheX	fails to tumble	28	mainly cytoplasmic/methyltransferase
42'	*cheW*	fails to tumble	12–15	cytoplasmic
	cheA	fails to tumble	76	mainly cytoplasmic
			66	cytoplasmic
43'	*cheC*	fails to tumble	—	—
98'	*cheD*	fails to tumble	65	—

behaviour depending upon conditions. The *cheA* gene is complex apparently encoding two distinct polypeptides, both essential for chemotaxis and differing only in that the larger has an additional 90-100 amino acids at the *N*-terminus. The smaller *cheA* product is thought to be cytoplasmic, whereas the larger polypeptide is associated with the cytoplasmic membrane.

Adaptation and deadaptation are closely correlated with the process of methylation and demethylation. The transducers are methylated by glutamyl residues to form carboxyl methyl esters. Donation of methyl groups from *S*-adenosylmethionine is mediated by a methyltransferase encoded by *cheX*. The *cheB* gene product has been shown to code for the methylesterase that demethylates the transducer. The function ascribed to *cheX* and *cheB* gene products correlates well with the mutant phenotypes. During adaptation, low methylation levels are associated with suppression of tumbling; *cheX* mutants are methyltransferase-deficient and express a smooth-swimming phenotype. High levels of transducer methylation are found in *cheB* mutants consistent with their tumbling phenotype.

When the transducers are labelled with radioactive methyl groups and analysed by sodium dodecyl sulphate (SDS)-polyacrylamide gel electrophoresis it is possible to resolve MCP into a series of 8–10 bands (ranging in M_r from 56,000 to 65,000). Lesions in two genes, *tar* and *tsr*, cause the loss of some, but not all of these polypeptides. The *tar⁻* mutants retain six of the bands, called MCPI, whilst *tsr⁻* mutants retain four of the bands (MCPII). Double mutants carrying both *tsr* and *tar* defects are unable to methylate any of the MCP bands. It is not clear why *tsr* and *tar* genes appear to produce multiple bands on SDS-polyacrylamide gels, but detailed analysis of the *tar* products reveals extensive regions of homology suggest-

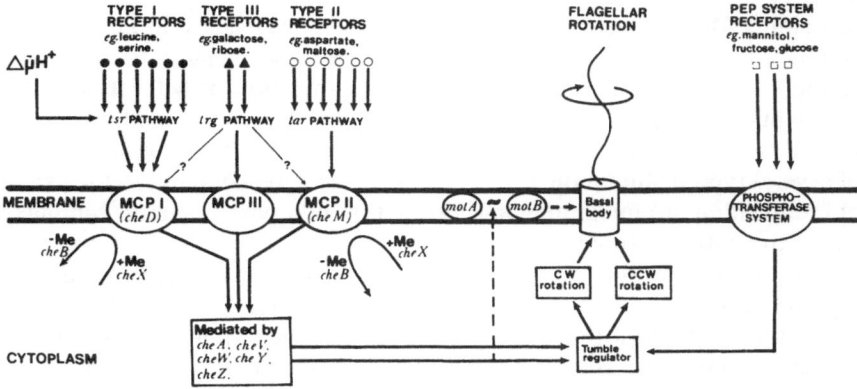

Figure 4.1: Schematic Representation of Information Flow During Chemotaxis in *E. coli.*
Chemotactic signals are processed in the periplasmic space and on the cell membrane, and
transmitted by the *tsr* pathway using MCPI, by the *tar* pathway using MCPII, or both. The MCP
undergoes methylation by the *cheX* methyltransferase or demethylation by the *cheB*
methylesterase. The information undergoes further processing by the other *che* products
present in the cytoplasm, resulting in its transmission to the tumble regulator. This serves to
control the frequency of changes in the direction of rotation of the flagellar shaft. The terminal
signal may act on both the rotor and the *mot* gene products. Methylation is thought to increase
the incidence of clockwise (CW) rotation and demethylation counterclockwise (CCW) rotation.

ing that some of the *tar* bands may be the result of the post-translational
modifications of a single polypeptide. Alternatively the multiplicity of
bands may reflect the degree of methylation. Two-dimensional electrophoresis has demonstrated that a methylation site present on the *tsr* and *tar*
gene products, is capable of accepting three or four methyl groups/
polypeptide. There is now strong experimental evidence for two distinct
methylation-dependent pathways for transmission of sensory stimuli from
the receptor to the flagellum in *E. coli,* one using the methylation of MCPI
and a second dependent on MCPII. A third pathway, that triggered by
galactose and ribose, is less well understood but may involve a third MCP
system (Figure 4.1).

Type I stimuli, which include chemoattraction to serine (high affinity)
and repulsion by leucine and indole, feed their signals primarily through
the pathway defined by MCPI. However under certain specified conditions,
type I receptors can utilise the MCPII pathway, although the signal then
has reversed polarity. In *tsr* mutants that have lost the MCPI pathway, type
I repellants act as attractants and vice versa. This is apparently achieved
using MCPII, since under these conditions type I repellants cause an

increase in the degree of MCPII methylation.

Type II stimuli, which includes aspartate, maltose and serine (low affinity) attraction and nickel and cobalt repulsion, rely on a pathway containing MCPII, since the response to these stimuli is lost in *tar*-deficient strains which lack MCPII. Addition of type II attractants to wild type strains of *E. coli* increases the degree of methylation of MCPII, whereas repellants cause a dramatic reduction in the degree of methylation of MCPII, with little change in MCPI methylation. On simultaneous addition of several type II stimuli the effect on chemotactic behaviour and the extent of methylation are integrated, suggesting that two type II stimuli can make use of the pathway simultaneously.

Non-motile mutants of *E. coli*, designated *cheM* and *cheD*, have been shown to possess a defect lying between the chemoreceptor and the general chemotaxis pathway. The gene products have been shown to be integral membrane proteins, *cheM* giving a series of polypeptides (M_r 60,000 to 63,000) as does *cheD* (M_r 63,000 - 66,000), and shown to correspond with the MCPs.

Chemoattraction to galactose and ribose (type III) may occur in *tsr*- or *tar*-deficient mutants but is diminished in *tsr*-*tar* double mutants. This suggests that type III stimulants utilise both MCPI and MCPII. However, since these compounds retain some residual effect in the double mutant it has been suggested that type III attractants additionally act through a third route, the *trg* pathway, that may or may not use MCPs I and II but does use a third protein, MCPIII.

The methylation of MCPs plays a central role in sensory adaptation, since if the methylation reaction is blocked then sensory adaptation is halted. A considerable body of additional evidence supports a link between methylation and adaptation. The kinetics of adaptation and MCP methylation appear similar; the attractant-induced increase in methylation and half-time of the reaction are directly proportional to the adaptation time, rates of adaptation and methylation are fast, whereas de-adaptation and demethylation are slow. Different stimuli presented simultaneously produce cumulative effects on both the methylation reaction and adaptation time. Maintenance of the adapted state and of attractant-induced methylation are both methionine-independent as is the process of de-adaptation and demethylation. The mechanisms that link methylation to adaptation are not yet fully defined, but the methylation reaction appears to alter the parameter that regulates the sensitivity of the organism to external chemical stimuli.

A number of membrane-permeant weak acids, e.g. acetate, propionate and benzoate, act not by binding to a chemoreceptor, but by shunting protons to the cytoplasm and hence lowering intracellular pH. Although these repellants do not bind to external chemoreceptors, transduction of the cytoplasmic pH signal involves the *tsr* system and MCPI, since *tsr*⁻ and methylesterase-deficient mutants give abnormal responses in the presence of weak acids. Additionally benzoate has been shown to induce demethylation of MCPI.

Many bacteria swim away from regions of high temperature. The thermoresponse in *E. coli* is inhibited by high serine or aspartate concentrations. This suggests that the *tsr* transducer may either serve as a thermosensor itself or the *tsr*-pathway transmits the signal from an unknown thermosensory device.

There is no evidence that any of the described methylation transducers are involved in chemotaxis responses to sugars transported by the phosphotransferase (PEP) system (Figure 4.1). The existence of further methylated transducers to account for these sugars has been postulated. Alternatively the phosphorylation that accompanies the uptake of the sugar may act as the transducer.

The flagellum does not contain all of the components necessary for flagellar rotation. Mutation of either the *motA* or *motB* genes leads to the production of paralysed flagella that appear otherwise normal. These two membrane proteins (M_r, *motA* = 33,000, *motB*=39,000) may provide the energisation necessary for flagellar motion.

The Tumble Regulator

Since Gram-negative bacteria swim straight with prolonged counterclockwise (CCW) rotation of the flagellum but tumble when they rotate in a clockwise (CW) fashion, the ultimate control of the chemotactic response resides in a hypothetical switch controlling the direction of flagellar rotation, known as the tumble regulator. The regulator controls the switch in the light of information received from the transducers. The *flaA* and *flaB* gene products are flagellar components active in the regulation of flagellar rotation. The observation that *cheW*− and *cheA*− mutants cannot tumble suggests that their products, which are cytoplasmic, may also be essential for reversal of the direction of rotation. In *Bacillus subtilis*, the intracellular divalent ion concentration may control the tumbling phenomena. It is possible to manipulate the intracellular levels of this ion by judicious use of a calcium ionophore and extracellular calcium. At low cytoplasmic calcium concentrations the cells swim smoothly, at slightly higher levels the organism exhibits mixed smooth swimming and tumbling and at high levels it tumbles incessantly. It has been suggested that the *B. subtilis* cell regulates the internal divalent ion concentration using transmembrane ion-fluxes and that internal calcium or magnesium concentration controls the direction of flagellar rotation.

Flagellar Structure

The predominant locomotory organelle of bacterium is the flagellum. This structure is only superficially similar to the organ of the same name present in eukaryotes. The bacterial flagellum consists of three clearly defined regions, the basal body embedded in the cytoplasmic membrane and a hook region which serves to connect the basal body to the long slender flexible shaft. The flagellum constitutes the 'H' (or Hauch) antigen. This antigen has been used in the classification and identification of flagellate bacteria, e.g. the Kaufmann-White scheme for the classification of Salmonellae.

The region between the distal end of the flagellum and the hook is known as the shaft or filament. Unsheathed filaments are approximately 20 nm in diameter and 10-20 μm long. In electron micrographs, i.e. when viewed as a flattened two-dimensional projection, they appear to possess a wave form of constant amplitude. The filament constitutes up to 95 per cent of the mass of the flagellum and consists of a single protein, flagellin. The molecular weight of the flagellin varies with species, e.g. *B. subtilis* M_r 33,000, *S. typhimurium* M_r 56,000 and *E. coli* M_r 60,000. The flagellin of *B. subtilis*, has been shown to be a single polypeptide of 304 amino acids (M_r 32,600). The flagellin has been sequenced; it lacks tryptophan and cysteine, and contains only a single tyrosine and two proline residues. The low cysteine content is believed necessary to protect the structure from conformational changes induced by differences in the redox potential of the environment. There is a random distribution of hydrophobic amino acids throughout the flagellin, but the distribution of charged amino acids creates three regions, the *N*-terminal is basic, the central region is acidic and the *C*-terminal weakly acidic. The amino acid sequence and the tertiary structure are crucial to the functioning of the organelle. A single amino acid substitution in the flagellin of *B. subtilis* 168 (valine for alanine at position 233) produces a non-sinusoidal flagellum incapable of movement. Less information is available on other flagellins but tryptophan and cysteine are invariably absent and the content of cyclic amino acids (Phe, Tyr, His and Pro) low. A degree of sequence homology has been reported between the flagellins of Gram-positive bacteria and this has been taken to suggest some evolutionary relationship. The rare amino acid ε-*N*-methyllysine is present in the flagellin of some Salmonella strains. The methylation of the lysine occurs after flagellin biosynthesis is completed. This methylation reaction is not essential to flagellar function, since mutants unable to perform this step are fully motile. Detailed study of the filament structure of *Proteus* species reveals that the flagellin monomers are wedge-shaped and exist in a regular array in strands parallel to the axis. Eight strands combine to form a hollow cylinder. The flagellin units are tilted at a 30° angle to the flagellum's axis and to the hollow centre (approximately 3nm wide). Biophysical measurements suggest that the bacterial flagellar filaments possess a high degree of flexural rigidity.

The filaments of *Bdellovibrio bacteriovorus* and *Vibrio metchnikovii* are sheathed. The core is of similar construction to that of unsheathed filaments and the sheath is thought to be an extension of a cell wall giving a much thicker filament (35 nm).

The section of flagellum between the filament and the basal body is termed the hook. Generally short and slightly curved it has a diameter slightly greater than the filament. The hook is less variable than the filament, usually 70-90 nm long. Like the filament it appears to consist of a single molecular species. The flagellin and hook proteins of Salmonellae have been isolated and shown to be antigenically distinct. The hook protein has a molecular weight of 43,000 and the flagellin 56,000. The hook protein has no methyl-lysine residues but contains more cyclic amino acids. Conditions that lead to the dissociation of the filament into flagellin subunits, e.g. low pH, urea or guanidine, have little effect on the hook. *Hag* mutants of *E. coli* are defective in flagellin biosynthesis but produce normal hooks and basal bodies. Although the flagellar filaments of Salmonellae show great antigenic diversity, i.e. a large number of H-antigens have been detected, they all share a common hook. The assembly of hook-subunits is controlled by the *flaE* gene, whch regulates hook length, since *fla⁻* mutants of *S. typhimurium* form polyhooks several μm long. The function of the hook is not fully understood. It has been suggested that the curvature of the hook in enteric bacteria orientates the filament, serving as a flexible coupling to the cell surface. This enables the lateral filaments to operate as a bundle along the long axis of the cell. The hook may also serve as a point of initiation for flagellin polymerisation.

The basal body is the most morphologically and functionally complex region of the bacterial flagellum. It serves to anchor the hook to the cell surface and provides the driving force for flagellar rotation. Since the flagellar apparatus serves as the final destination of the chemotactic signal certain of the components involved in chemotaxis may also be located in the basal body. Dissociation of the basal structure reveals that it contains at least ten distinct polypeptides. The hook is attached to a hollow rod that spans the bacterial envelope. The hook of Gram-positive bacteria is generally longer than that found in Gram-negative organisms but the rod is shorter. The rod of·Gram-negative bacteria, e.g. *E. coli*, ends in a disc-like structure, the M(membrane)-ring, which is embedded in the cytoplasmic membrane (Figure 4.2). Closely associated with the M-ring is found a second disc, the S(supra-membrane) ring. It is thought that the M-ring serves as a rotor and the S-ring as a stator during flagellar rotation (see later). The P (peptidoglycan) and L (lipopolysaccharide)-rings are tightly associated with the outer membrane of *E. coli*. Since P- and L-rings are not present in Gram-positive bacteria for example *B. subtilis* (Figure 4.3) it has been suggested that they may not be directly involved in flagellar rotation and serve as a bearing permitting rotation of the rod in the relatively mobile outer membrane. In *B. subtilis* the S-ring may be attached to the innermost layer of the peptidoglycan. The grommet observed in the Gram-

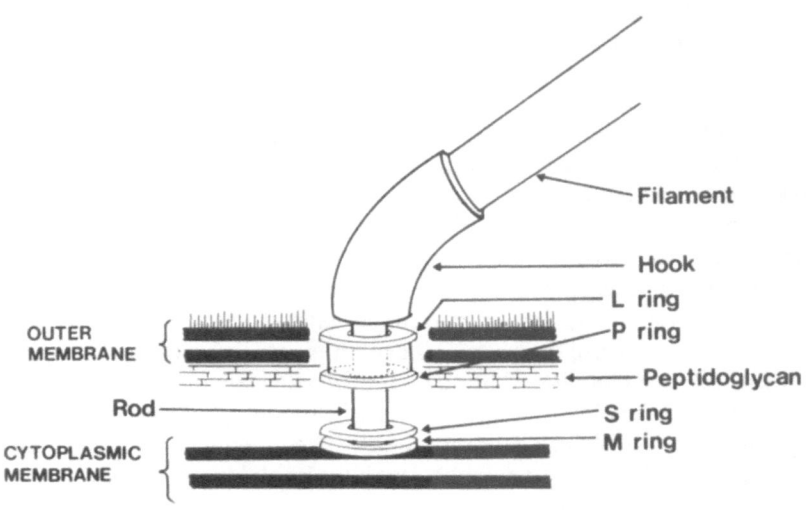

Figure 4.2: Flagellar Apparatus of Gram-negative Bacteria

positive *Clostridium sporogenes* may also act as a bearing. It is possible to detach the rings from purified rod preparations using homogenisation. Rotational analysis of isolated discs suggests that each consists of 16 wedge-shaped protein subunits. The basal body is also involved in the assembly of the flagellum. During flagellar morphogenesis a certain degree of localised peptidoglycan hydrolysis is required for assembly and insertion of rings into the envelope.

The mechanism of flagellum biosynthesis has been extensively studied in *E. coli* and *S. typhimurium*, but these findings have also been generally applicable to other species, e.g. *Proteus mirabilis* and *B. subtilis.* The assembly and function of the flagellum is controlled by approximately 20 genes, clustered at the 43' position of the *E. coli* chromosome (close to the *che* and *tar* loci). The primary sequence of flagellin is determined by the *hag* (H-antigen) gene. The proteins of the hook and basal body are the products of the *fla* (flagellar) genes. Other *fla* genes regulate flagella production, since their mutation leads to changes in the number of flagella/cell. Additionally other genes (termed *mot*) are believed to control flagellar rotation. The *mot* gene product of *E. coli* has been identified as a 30,000 molecular weight polypeptide, present in the cytoplasmic membrane, which is essential for flagellar rotation. Mutation of this gene leads to flagellar paralysis even though the flagella appear normal. Many of the genes controlling motility in Gram-negative bacteria are grouped in multi-

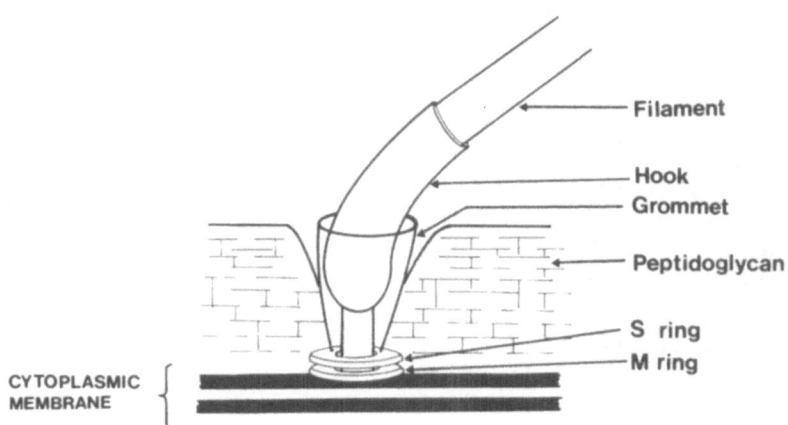

Figure 4.3: Flagellar Apparatus of Gram-positive Bacteria

cistronic units, e.g. *flaE, flaO, flaC* and *flaB* are transcribed *en bloc.* A given Salmonellae serotype may undergo changes in flagellar antigenicity known as 'phase change' in which the organism produces flagellin of one of two types. Each flagellin is the product of a separate structural gene *hag 1* and *hag 2*. Associated with the *hag 2* gene is a regulator gene *rh 1*, which produces a repressor for the transcription of *hag 1*. The *hag 2 rh 1*, which has been shown to oscillate between active and inactive states corresponding with the observed phase changes. The mechanism by which the cell determines the site of flagellar insertion is not known.

In *E. coli* the biosynthesis of flagella is subject to glucose catabolite repression. In the presence of glucose the rate of flagellar synthesis is decreased but on exhaustion or removal of the carbohydrate synthesis recommences. Catabolite repression is mediated by 3':5'-cyclic AMP (cAMP). Addition of cAMP to repressed cells will stimulate flagellar synthesis; cAMP reacts with cAMP-receptor protein and together modulate *flaT* which in turn regulates the *hag* genes responsible for flagellar biosynthesis. It is presumed that the respective enzyme II mediates catabolite repression of flagellar synthesis.

Regeneration of flagella after mechanical deflagellation is rapid. In *E. coli* the flagellar filament grows at about 0.2 μm min^{-1}, the rate gradually slowing as the structure reaches maximum length. Flagellin monomers spontaneously aggregate without enzymic mediation. On treatment with urea, guanidine, detergents or exposure to extreme pH, flagellar filaments dissociate into their constituent monomers. The molecules may reassociate to form filament-like structures practically indistinguishable from the parent structure, excepting for greater length. *In vitro* polymerisation experiments using *Salmonella* flagellin suggest that a monomer binding to the distal end of the filament, undergoes a conformation change and then serves as the nucleus for the polymerisation of the next monomer. This conformational change of flagellins on assembly confers structural polarity on the filament confining the assembly to the distal end. Radio-labelling experiments have shown that during flagellar growth in *B. subtilis* cells the filaments elongate by polymerisation at the distal end only. Since free flagellin cannot be detected outside the bacterium it is presumed that the monomers are transported through the hollow core of the filament until they reach the tip where their powers of self-assembly ensure correct orientation in the growing tip.

Flagellar Motion

An important concept in understanding bacterial chemotaxis is the phenomenon known as adaptation. Following delivery of the stimulus, even though the stimulus is maintained, there will be a decline in the observed response. This is termed adaptation. Upon removal of the stimulus the bacterium regains its sensitivity and will again respond if the signal is repeated (de-adaptation). The behaviour of the organism is determined by the balance between the two processes: stimulation of the receptor initiating a response and the slower adaptation which extinguishes the response.

Locomotion is mediated by the rotation of the flagella filament. Movement of the flagellum cannot be followed under the light microscope; however, if anti-flagellin antibody is used to link latex beads to the filament then a chain of attached beads can be observed revolving in unison about a common axis. Antibody molecules can also be used to anchor catabolite-repressed *E. coli* cells which only possess a single flagellum to a glass slide. Cells tethered by antibody are seen to rotate around the point of attachment. The flagellar filaments rotate at about 50 revolutions per second whereas, because of viscous drag, the cell rotates approximately ten times slower in the opposite direction, in accordance with Newton's laws of motion. Flagellar rotation generates a thrust on the bacterium, which moves forward with the flagellum trailing behind, i.e. the rotating flagellum drives the cell forward like a ship's propellor (Figure 4.4.).

Peritrichous bacteria, like *E. coli* and *S. typhimurium*, bear 8–10 flagella positioned randomly on the cell surface. Rotation of these flagella results in the filaments turning in a co-ordinated manner as a helical bundle (Figure 4.5). The ability to form a flagellar bundle is made possible by the flexible nature of the hook region acting as a universal joint. The flagellar filaments of *E. coli* exist as semi-rigid helices with a left-hand pitch. During counter clockwise (CCW) rotation the filaments form a compact bundle capable of

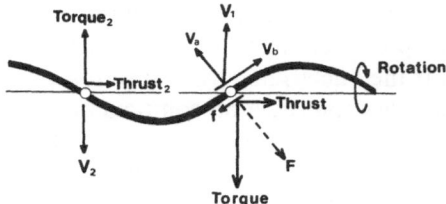

Figure 4.4: Analysis of Forces Acting upon a Rotating flagellum (after Berg 1975). The velocity (V) of a point on the filament can be resolved into a perpendicular component (V_a) and one parallel to the filaments' axis (V_b). Viscous drag will be proportional to the velocity, but will be greater for motion perpendicular to the filament than for motion parallel to it. Drag perpendicular (F) and parallel (f) to the flagellum are resolvable into components that contribute to the torque and thrust motions. Similar analysis of a half wavelength distant shows that equivalent thrusts (Thrust$_2$ and Torque$_2$) are developed, but in the opposite direction. Forces perpendicular to the axis, but acting in opposite directions will tend to rotate the cell body, while forces parallel to the axis will propel the bacterium through its environment.

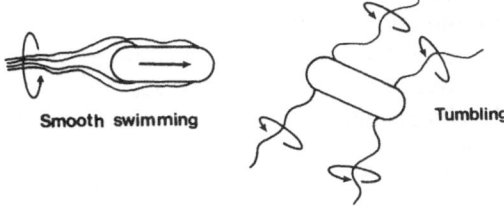

Figure 4.5: Locomotion in Peritrichous Bacteria, e.g. *E. coli* and *S. typhimurium*

acting together to push the organism through the medium in a straight line, with a velocity of $25\mu m\,S^{-1}$. If the direction of rotation is reversed (clockwise; CW rotation) the bundle filaments are then thrown apart causing the tumbling response (CW and CCW rotation being defined as if the observer were looking along the filament axis towards the bacterium). The random uncoordinated movements of tumbling are caused by gross conformational changes in the filaments, induced by the reversal torque of clockwise rotation. The filaments, under these conditions, exhibit right-handed helicity, reduced amplitude and wavelength, a state in which bundle formation cannot occur. Each tumbling episode results in a nearly random re-orientation of the direction of swimming, ensuring little correlation between the direction of successive runs. A return to CCW rotation leads to the relaxation of the helix to the left-handed condition, the reformation of the bundles and the resumption of swimming (Figure 4.5). Bacterial behaviour in an isotrophic environment is characterised by periods of smooth swimming, interspersed with brief episodes of tumbling.

The tumble is the principal mechanism by which *E. coli* alters the direction of movement. If the bacterium swims towards an increasing attractant concentration the sensory transduction components detect the increase and the probability of tumbling is decreased. This results in longer periods of swimming in the direction of increasing attractant concentration and the bacteria gradually accumulate in the region of highest attractant levels. The converse applies during negative chemotaxis, when the bacteria swim towards regions of decreasing repellant concentration the tumbling frequency decreases. The tumbling response offers a simple, but extremely effective, control mechanism. This simplicity makes the mechanics of the process uncomplicated. The organism need only be concerned with regulating the transition between two behavioural states, the CW and CCW rotation of the flagellum. The use of a temporal sensing mechanism to monitor the spatial gradient permits amplification of the stimulus.

In polar flagellated bacteria, e.g. *Chromatium*, *Pseudomonas aeruginosa* and *Rhodospirillum*, the result of changes in the direction of flagellar rotation is less dramatic but no less effective than in *E. coli*. On the reception of the signal, the flagellar rotor reverses and the organism swims off in the opposite direction (Figure 4.6).

Bacteria of the genus *Spirillum* exhibit a spiral morphology (the cell has a distinct right-handed helicity) and possesses non-helical tufts of polar flagella. These flagella produce cones of rotation at one or both ends of the cell causing rotation of the helical cell. Reversal of the direction of flagellar rotation causes simultaneous inversion of both flagellar bundles (rather like an umbrella being turned inside out in a strong wind) (Figure 4.7). Similar behaviour has been reported for bipolar flagellated bacteria and *Halobacterium*, which bears bipolar multiple flagella.

The filaments of the flagellar bundle of *E. coli* all rotate in the same direction. Tumbling requires simultaneous reversal of all the flagellar rotors. Similarly the bundles of *Spirillum* although at opposite poles of the cell exhibit synchronous changes in rotational direction. These properties suggest the existence of a central co-ordinating system, possibly an electrochemical mechanism, similar to that of the ciliate protozoa.

Figure 4.6: Locomotion in Polar Flagellated Bacteria, e.g. *Pseudomonas*

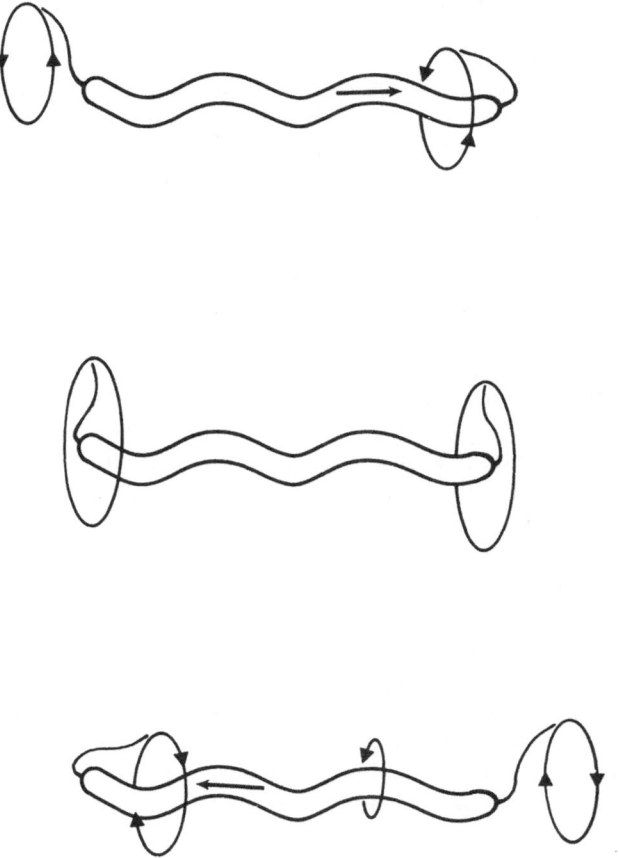

Figure 4.7: Reversal of Rotation in *Spirillum volutans*. This organism bears two polar tufts of flagella.

The driving force of eukaryote flagella and muscle contraction is the hydrolysis of ATP. However, ATP-hydrolysis is not neccesary for flagellar rotation in bacteria. *E. coli* AN120 lacks an effective ATPase, but has a functional respiratory chain. This strain cannot synthesise ATP by oxidative phosphorylation but can carry out substrate level phosphorylations (glycolysis) leading to the formation of ATP. *E. coli* AN120 is highly motile in the presence of oxygen but under anaerobic conditions movement ceases. Depletion of cellular ATP reserves fails to inhibit the mobility of the organism if oxygen and oxidisable substrate are present. Uncouplers of oxidative phosphorylation immediately arrest flagellar motion. These results and others were interpreted as demonstrating the involvement of '~' the intermediate of oxydative phosphorylation as the driving force of flagellar rotation. The chemiosmotic hypothesis explains '~' in terms of an electrochemical gradient of protons ($\Delta\bar{\mu}H^+$) existing across the cell membrane. The passage of protons down such a gradient can be made to do work, e.g. active transport. It is believed that the torque necessary for flagellar rotation is generated as protons are channelled between S- and M-rings of the basal body. No current model fully explains how the basal body can rotate in the plane of the membrane, while permitting the passage of protons down an electrochemical gradient. Hydrodynamic estimates of power dissipation and of $\Delta\bar{\mu}H^+$ can be used to determine the number of protons required. If P joules of hydrodynamic work are being performed each second, by a rotor running at ΩHz, using protons bearing E joules of free energy each at least P/E ΩHz protons will be required per revolution (assuming 100 per cent energy transfer). If $\Delta\bar{\mu}H^+$ is − 200 mV, a tethered cell radium 1 μm and a speed of rotation of 7.5 Hz would require just over 200 protons/revolution. More protons would be required if the cell were rotating more slowly due to say increased viscosity, e.g. a cell rotating at 2.3 Hz would require 720 protons/revolution.

One possible model suggests that the M-ring is encircled by amino groups, one of which is accessible to an upper proton conducting path (Figure 4.8). Anionic carboxyl groups are fixed to the membrane at the entrance to a lower proton conducting path. Protonation of the amino group leads to electrostatic attraction between the amino and carboxyl groups. The distance between the two proton half channels must be equal to that between adjacent amino groups. Hence rotation of the M-ring results in the next amino group being transferred to the upper proton path. Protonation of the carboxyl group by the protonated amino group results in

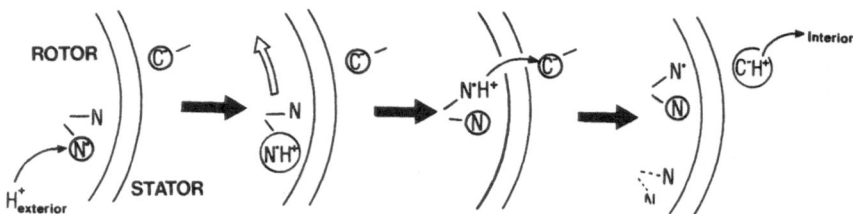

Figure 4.8: Model of Proton-driven Rotation. This model postulates a series of single steps to generate rotation. Amino-groups (N) exist at intervals around the rotor and carboxyl-groups (C) on the stator in the membrane. A proton (H$^+$) from the environment enters a port above the rotor and protonates the amino-group (N*). Electrostatic attraction between this group and the carboxyl-group on the stator rotates the rotor to bring together N*H$^+$ and C$^-$, the proton transfers to the carboxyl group and finally enters the cytoplasm. The rotation brings the next amino-group under the proton port, ready to repeat the process.

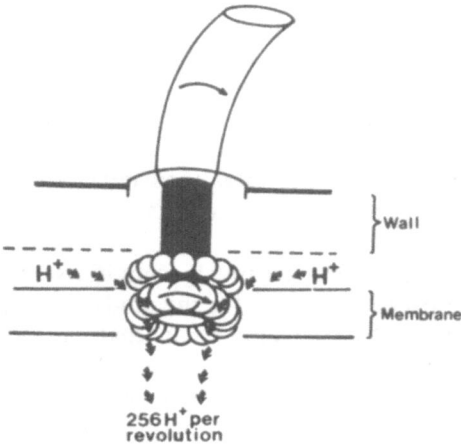

Figure 4.9: Model of Flagellar Rotation (after Berg, 1975). The lower M-ring consists of 16 protein sub-units, opposed to a similar fixed fing, the S-ring. Passage of a proton, in response to the protonmotive force, through each M-ring sub-unit causes the rotation of the flagellum through one-sixteenth of a turn, i.e. 256 protons would be consumed in each complete revolution.

extrusion of a proton into the cytoplasm along the lower path. This model requires at least 200 amino groups at approximately 30 nm intervals around the M-ring and the flagella would rotate by a ratchet-like mechanism of 200 distinct steps. However, it is suggested that several protons must pass through the motor simultaneously, since this would ensure unidirectional rotation and lower the activation energy. To permit reversal of the flagellum one must also postulate a second proton channel symmetrical to the first. The direction of rotation would then be determined by the opening and closing of the appropriate channel under the direction of the chemotactic signal. An alternative, and somewhat less complex mechanism has also been suggested. The M and S rings of the basal body consist of 16 subunits and a plausible stoichiometry has been invoked involving one ion-channel per subunit per step to the next channel, i.e. $16 \times 16 = 256$ protons/revolution in a vernier arrangement (Figure 4.9). The protons presumably flow through the subunits of the M-ring generating force against the adjacent ring.

Attempts have been made to determine how the magnitude of $\Delta\bar{\mu}H^+$ affects flagellar rotation. Upon complete dissipation of $\Delta\bar{\mu}H^+$ the filament is not free to rotate, appearing rigidly engaged, almost as if it were locked in gear. The flagellum does not appear capable of powered rotation until a threshold of about 25 mV is passed. The rate of flagellar rotation then increases directly in proportion to $\Delta\bar{\mu}H^+$ but reaches maximum speed at a relatively low $\Delta\bar{\mu}$ (about 100 mV). The motor does not appear to discriminate between a negative or positive potential, the direction of rotation being apparently independent of whether protons are entering or leaving the cell. Changes in $\Delta\bar{\mu}H^+$ towards zero increase the tendency towards CCW rotation.

Certain components of the ion-conducting pathway have been shown to correlate with the application of chemotactic stimuli and to be missing in particular *mot⁻* mutants, possessing paralysed intact flagella. It has been suggested that the mot-products may serve as 'ion-gates' or be involved directly in energy transduction.

Bacterial chemotaxis has received much scientific study, as the system was thought to be an excellent model for the study of simple behavioural responses. It has considerable advantages in that the responses can be easily quantified and subjected to biochemical and genetical analysis. Although much is now known about the bacterial response and flagellar motion, many parts of the system remain obscure. However these studies have clearly demonstrated that bacterial chemotaxis is not a 'simple' system at all, but a highly complex integrated system of sensors, transducers and motor components, whose functional inter-relationships we are only beginning to understand.

Non-flagellar Locomotion

It would be wrong to leave the impression that direct flagellar rotation is the only or even the most important locomotory system in bacteria. All spirochaetes show active movement, however they do not appear to possess external locomotory organelles. Running the length of the cell is an axial filament composed of between 2 and 100 fibrils, each fibril resembling a bacterial flagellum which has become internalised. The fibrils are situated within the periplasmic space and radiate from each pole, overlapping in the centre of the cell. The fibrils originate from a basal body inserted into the cytoplasmic membrane at the cell pole. Attached to the basal body are hook and filament regions, all of which show great similarity to the equivalent structures present in bacterial flagella. The fibrils are made up of protein subunits (M_r 37,000) that resemble flagellin in composition. The fibrils may be ensheathed in other layers.

Berg has suggested spirochaete motility is achieved by rotation of the axial filaments. If the external layers of the organism are considered to be rigid, and loose fitting, then the rotation of the axial filaments would cause the outer layers to slip relative to the protoplasmic cylinder. This will force the cell to rotate about its helical axis in the opposite direction and drive the organism through the medium in a corkscrew fashion. The torque required to rotate the cell about its helical axis would be provided by circumferential viscous shear due to the roll of the external layers (Figure 4.10).

Slender spirochaetes often exhibit a flexing motion and this may be the result of a twisting of the protoplasmic cylinder, produced in reaction to the

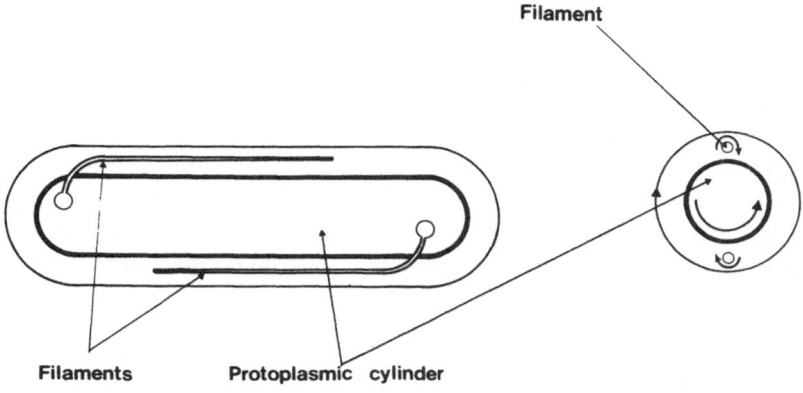

Figure 4.10: Representation of Locomotory Filaments of Spirochaetes (after Berg, 1976). In reality the periplasmic cylinder is helical not straight as shown. Only one filament per pole is shown, although at least two and often more filaments are present. The rotation of the filaments cause the cylinder, together with the filaments, to rotate within the loosely fitting flexible external sheath. Since the external sheath is free it also rotates, but in the opposite direction.

torque exerted by rotation of the filaments. The curvature and torsion is most pronounced in thin cells. Rotation of the filaments in the same directions cause curling motions of the cell poles in the same direction; if they rotate in opposition the cell will tend to flex.

Some spirochaetes exhibit a creeping motility. If the organism approaches a solid surface the external viscous shear will be largest when the cell and the surface are closest. If the protoplasmic cylinder is long and irregular, it may not be free to rotate, and the roll of the exterior layers cause the cell to slide in a direction nearly parallel to the local helical axis, i.e. producing a creeping-type motility.

Internalised flagella may not be unique to spirochaetes. Similar structures have been identified in the blue-green alga, *Oscillatoria princeps*, which possesses a parallel array of very fine fibrils (5–8 nm thick) under the cell surface. However the precise basis of locomotion in this species is not fully understood.

An additional type of movement along side surfaces, termed gliding movement, is present in some mycoplasmas, all species of the order Cytophagales and fruiting myxobacteria. These organisms do not appear to possess external locomotory appendages or internal filaments, but it is thought that since they move across solid surfaces, the phenomenon is surface related.

Non-flagellar Surface Appendages; Fimbriae and Pili

Many bacteria possess surface appendages in addition to the flagella. These are usually seen under the electron microscope as thin, non-sinusoidal filaments radiating from the organism and are given the name fimbriae. The common fimbriae are smaller and more numerous than flagella and occur widely in Gram-negative bacteria and in a single Gram-positive group, the Corynebacteria. These fimbriae are believed to play an important role in the attachment of bacteria to surfaces. A special class of fimbriae, the sex pili, are restricted to Gram-negative bacteria, being present in low numbers per cell and appearing much longer and wider than the common fimbriae. The formation of a sex pilus, depends upon the presence within the organism of a sex factor or a plasmid bearing the genes for pilus production. The sex pilus is involved in the transfer of genetic material during bacterial mating.

Fimbriae of Gram-negative Bacteria

A single bacterium may bear between 100 to 1,000 fimbriae, varying in length (between 0.2 and 20 μm) and in width (3–14 nm). They are distributed evenly around the cell surface. Fimbriation is a reversible trait and is most likely determined by chromosomal gene(s). Bacteria are subject to mutation leading to sudden, complete and often reversible loss of fimbriae. The rate at which fimbriae are lost is high (may approach 10^{-2}/cell division) but the reversion rate is much lower. The rate of fimbrial loss is influenced by environmental factors, e.g. temperature, oxygen tension or agitation. It has been suggested that a mutator gene controls the degree of fimbriation and this is subject to environmental control. For full fimbrial expression in enterobacteria the organisms are grown in serial cultivation in static nutrient-rich broth at 37°C. Exponential phase growth, or non-selective growth on solid media leads to a marked reduction in the number of fimbriae. This phenomenon also occurs in *Vibrio* and *Neisseria* species, but in meningococci growth of the organism on solid media does not reduce the number of fimbriae. Pseudomonads generally bear more fimbriae during log phase and this number declines as the culture enters stationary phase.

Duguid used the variation in morphology and haemagglutinating properties of fimbriae from Gram-negative bacteria to develop a classification system.

Type 1 (Common) Fimbriae (Figure 4.11). Approximately 400 of these organelles, about 7 nm wide and 2 μm long are peritrichously located. These fimbriae are responsible for the adhesive properties of the strain, in particular the ability to agglutinate red blood cells.

Enterobacteria with type 1 fimbriae adhere to fungal, plant and animal

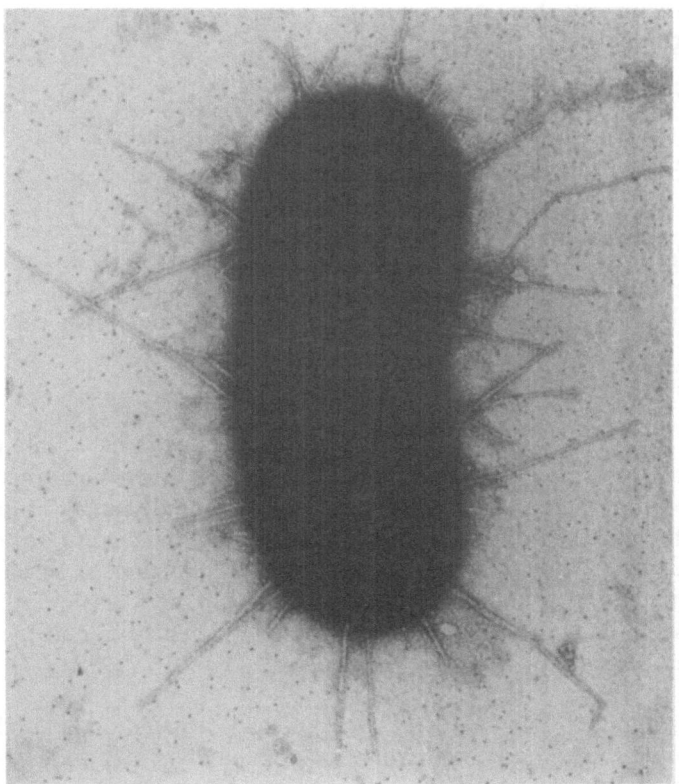

Figure 4.11: Fimbriae in *E. coli*. (Photograph by permission of Iwah, T., Abe, Y. and Tsuchiya, K. (1982) *J. Med. Microbiol., 15,* 303.)

cells. Guinea-pig, fowl and horse erythrocytes are strongly agglutinated by these fimbriae and since this interaction is inhibited by the sugar, D-mannose or methyl-α-D-mannoside, these adhesins are termed mannose-sensitive (MS). Type 1 fimbriae also confer the ability to form pellicles of adherent bacteria on the surface of a static broth culture. Type 1 fimbriae have been described in *E. coli, Salmonella, Klebsiella, Serratia* and *Enterobacter.* The widespread distribution of type 1 fimbriae among saprophytic, commensal and pathogenic bacterial groups poses difficulties in the understanding of their significance. Although much evidence exists to indicate that adhesion allows initial colonisation, a crucial step in mutualism or pathogenic attack, the role of the organelle in saprophytic bacteria remains less clear. Many enterotoxigenic strains of *E. coli* isolated from humans, piglets and calves have been shown to possess surface organelles (known as colonisation factor, K88 and K99 respectively) morphologically identical with type 1 fimbriae, but which mediate mannose-resistant (MR) haemagglutination.

Type 2 Fimbriae. These appear morphologically identical to those in type 1 and are present on a few *Salmonella* species, but differ in that they lack adhesive and haemagglutinating properties.

Type 3 (Thin Type) Fimbriae. Filaments of this type are restricted to *Klebsiella* and *Serratia marcescens.* They differ from type 1 fimbriae in that they are thinner (4.8 nm in diameter) and more numerous. They are characterised by a strong MR adhesion to fungal and plant cells and to glass surfaces. Type 3 fimbriae are incapable of binding to erythrocytes or animal cells, unless pretreated with tannin.

Type 4 Fimbriae. Many *Proteus* spp. produce extremely thin (4 nm), peritrichously arranged filaments that confer MR haemagglutinating and adhesive properties on the cells. These fimbriae predominate in exponential broth cultures and may account for the ability of *P. mirabilis* to cause pyelonephritis. Unlike type 1 fimbriae those belonging to this class are most active in the haemagglutination of sheep and fowl erythrocytes. Similar small diameter MR fimbriae have been detected in non-pathogenic *Neisseria* species.

Type 5 Fimbriae. These are the MS, monopolar contractile fimbriae of *Pseudomonas echinoides,* which are approximately 5 nm wide, 2–10 μm long, few in numbers and adhere to sheep erythrocytes. Similar structures have been observed in *Agrobacterium* and *Rhizobium* species. These structures pull bacteria together, by contraction, into star-form cell clusters. These fimbriae serve in recognition, contact and the irreversible joining of competent cells but it is thought that they do not serve as bridges for the direct transfer of nucleic acids.

Fimbriae of Gram-positive Bacteria

Amongst Gram-positive bacteria a single genus, *Corynebacterium*, has been shown to possess fimbriae. Cells of *Corynebacterium renale*, responsible for bovine pyelonephritis and cystitis, produce characteristic bundles (0.4 × 10 μm), composed of a number of individual fimbriae each 3 nm in diameter. These appear superficially very similar to the type 4 fimbriae of Gram-negative bacteria. Several studies have indicated that the Corynebacteria fimbriae are also important in adhesion and pathogenesis.

Table 4.4: Properties of F and I Pili on *E. coli*

Type of pili	F	I
Diameter (nm)	7.5–13.5	6-12
Maximum length (µm)	20	2
Plasmid determinant	F (or F-like e.g. R100-1, Fc-lac)	I (or I-like drug resistance factors)
Specific phages observed	F, f2 MS2	If1 If2
Pili in presence of acridine orange	None	present

Sex Pili of Enteric Bacteria

Pili are filamentous appendages present on the surface of male (F$^+$ and Hfr) strains of *E. coli* and other enterobacteria. The possession of pili is associated with the presence in the strain of a transmissible extrachromosomal element called a plasmid. Only those plasmids capable of promoting chromosomal transfer contain genes directing synthesis of pili. The plasmids may also confer antibiotic resistance, the ability to produce bactericidal antibiotics called colicins or the utilisation of obscure metabolites.

In *E. coli* two distinct groups of sex pili can be identified (Table 4.4). One group includes all pili that resemble the F-(fertility) pilus. Such pili are determined by the F-like plasmids, including the fertility factor itself, the drug resistance plasmid R100.1 and Fc-lac. Strains containing the colicinogenic plasmic *Col*-I produce a morphologically distinct pilus. The I-pilus (and similar structures present on strains containing I-like drug resistance plasmids) serves as a specific receptor for certain phages, e.g. the filamentous DNA phages If1 and If2. Similarly the F-pili are specific receptors for RNA phages like f2 and MS2 and filamentous DNA phages M13 and f1. Both F and I pili can occur simultaneously on a single cell, but then each pilus functions independently under the control of its respective plasmid. Pili are also specified by antibiotic resistant plasmids belonging to the incompatibility groups N, P, T, W and X, but these are distinct from I and F pili.

Sex pili can be readily distinguished from fimbriae, even in organisms bearing both fimbriae and pili. The sex pili are present in low numbers per bacterium (usually between 1 and 10). The pili are longer and wider than the fimbriae (6–13.5 nm wide and up to 20 µm long), exhibit a distinct axial hole (2.25 nm diameter) and often bear a terminal knob (15–80 nm diameter), presumably the membranal base of the filament. The proteins that make up pili and fimbriae are biochemically and antigenically distinct.

Table 4.5: Chemical Properties of Fimbrial Subunits

Species	Molecular weight of fimbrial subunit ($\times 10^{-3}$)	Carbohydrate content (%)	Isoelectric point
Escherichia coli (type 1)	16.6–17.5	0	4.5–5.1 (3 bands)
Escherichia coli K88	25	0.6	—
Escherichia coli K99	major subunit 22.5 minor subunit 29.5	1	10.1
Neisseria gonorrhoeae	19–22	1-3	5.3
Pseudomonas aeruginosa (K)	18	0	3.9
Corynebacterium renale	19–19.4	0	4.3

The absorption of male specific phages readily serves to identify the sex pili even in the most heavily fimbriate strains.

Pili are not thought to be important in bacteria-mammalian cell inter-actions but instead play an essential role in the bringing together of bacteria and subsequent transfer of genetic material between competent cells during conjugation. Bacterial cells containing conjugative plasmids are unable to transfer DNA if the plasmid contains defective pili-specifying genes. Similarly removal of the pilus by mechanical shearing produces bacteria unable to serve as gene donors until new pili develop on the cell surface. The attachment of antibody molecules or specific phages to the sex pili also inhibits conjugation.

Structure of Fimbriae and Pili

The isolation of non-flagellar appendages is relatively straightforward. The fimbriae are brittle and can be sheared by mechanical agitation then harvested by differential centrifugation.

Type 1 fimbriae of *E. coli* are exclusively protein, composed of a subunit of molecular weight 16,000–17,000 depending upon strain (Table 4.5). Analysis of the amino acid content has revealed a high proportion of non-polar residues. The adherence properties of type 1 fimbriae have been attributed to their essentially hydrophobic nature. Treatments known to disrupt hydrogen bonding cause the release of the fimbrial subunits, suggesting that these forces are involved in subunit–subunit interaction. Studies with the electron microscope, by X-ray diffraction and crystallo-graphy indicate that type 1 fimbriae are rigid, right-hand helices containing an axial hole 2–2.5 nm in diameter, a 2.4 nm pitch and a helical repeat distance of 2.32 nm. Like the flagellar subunits the fimbrial-subunits have the power of spontaneous reaggregation under appropriate environmental conditions.

The fimbrial subunits of enterotoxigenic *E. coli* are longer than those of common fimbriae. K99 fimbriae are composed of two subunit species; a major (M_r 22,000) and a minor subunit (M_r 29,500). K88 fimbriae exist as several antigenic variants, all bear the common antigen factor designated *a*, and in addition may bear either *b*, *c* or *d*. It has been shown that the K88*ab* plasmid specifies four polypeptides (M_r 17,000, 26,000, 27,000 and 81,000), the largest of which is an outer membrane protein. In the absence of the 81,000 molecular weight polypeptide, the newly synthesised smaller polypeptides accumulate in the periplasmic space and it has been suggested that the larger polypeptide is instrumental in the incorporation of fimbrial subunit precursors into the developing organelle. Both K88 and K99 fimbriae contain a high proportion of hydrophobic amino acids and bear three residues of the rare amino acid hydroxylysine.

The polar fimbriae of *Pseudomonas saeruginosa* K serve as receptors for a series of six bacteriophages and are built up from subunits of molecular weight 17,800 ($\pm$300). The serologically distinct polar fimbriae of *P. aeruginosa* O possess smaller subunits (M_r 15,500). Analysis of the amino acids of *P. aeruginosa* K fimbriae showed a large number of hydrophobic residues and a degree of helicity. The fimbriae of *Neisseria gonorrhoeae* exhibit antigenic variation, the subunit molecular weight α from antigenically distinct gonococcal fimbriae from as low as 17,000 to 21,000. Examination of the binding of isolated gonococcal fimbriae to human cells suggests that the point of attachment is heat-sensitive and may contain tryptophan residues, since modification of this moiety destroys the adherence property. The subunit consists of predominantly hydrophobic amino acids.

Analysis of the fimbriae present on the Gram-positive organism *Corynebacterium renale* shows them to consist entirely of protein with a subunit molecular weight of 19,000–19,400, again with a high content of hydrophobic residues.

The detailed study of the structure of the sex pilus of enteric bacteria is made difficult by their low number per bacterial cell. Their purification would be facilitated if strains could be freed of flagella and fimbriae. Unfortunately attempts to produce non-fimbriate mutants from non-flagellate male strains of *E. coli* have in the main been frustrated. However by harvesting the bacteria at the optimum time for pilus production and removing all filamentous appendages by mechanical shearing a pilus-rich mixture can be prepared. Subsequent differential centrifugation, equilibrium density centrifugation and iso-electric focusing of this mixture produces a pure protein fraction capable of binding male-specific phages and presumed to be the pilus protein termed pilin. Purified F-pili are composed of a single subunit type, consisting of a single polypeptide, a single D-glucose residue and two phosphate groups. Each subunit has a molecular weight of about 11,400. F-, I- and R-pilins are antigenically and biochemically distinct. Crystallographical studies have indicated that the F-pilin subunits are helically arranged around a central hole (diameter 2 nm) forming a flexible hollow tube 8 nm in diameter. It is suggested that significant differences exist between the tip and sides of the F-pilus, reflecting the role of the tip in establishing intimate contact between the mating partners. The phage F1 only absorbs to the tip of the F-pilus, whereas F2 and R17 preferentially bind to the sides.

Further Reading

Adler, J. 'The Sensing of Chemicals by Bacteria', *Scientific American* (1976), *234*, 40-7

Beachey, E.H. (ed.) *Bacterial Adherence* (Chapman and Hall, London, 1981)

Berg, H.C. 'How Bacteria Swim', *Scientific American* (1975), *233*, 36-44

Berg, H.C. 'How Spirochetes May Swim', *Journal of Theoretical Biology* (1976), *56*, 269-73

Berg, H.C., Manson, M.D. and Conley, M.P. 'Dynamics and Energetics of Flagellar Rotation in Bacteria', Symposium of the Society for Experimental Biology, No. XXXV. *Prokaryotic and Eukaryotic Flagella*, eds W.B. Amos and J.G. Duckett (Cambridge University Press, 1982), pp. 1–31

Boyd, A. and Simon, M. 'Bacterial Chemotaxis', *Annual Reviews of Physiology* (1982), *44*, 401-517

Costerton, J.W., Geasey, G.C. and Cheng, K.J. 'How Bacteria Stick', *Scientific American* (1978), *238*, 86-95

Doetsch, R.N. and Sjoblad, R.D. 'Flagellar Structure and Function in Eubacteria', *Annual Reviews of Microbiology* (1980), *34*, 69–108

Ellwood, D.C., Melling, J. and Rutter, P. (eds) *Adhesion of Microorganisms to Surfaces* (Academic Press, London, 1980)

Hazelbauer, G.L. (ed.) *Taxis and Behaviour: Elementary Sensory Systems* (Chapman and Hall, London, 1980)

Hazelbauer, G.L. and Parkinson, S.J. 'Bacterial Chemotaxis', In J.L. Reissig (ed.) *Microbial Interactions*, (Chapman and Hall, London, 1977) pp. 61–98

Heinrichsen, J. 'Twitching Motility', *Annual Reviews of Microbiology* (1983), *37*, 81–93

MacNab, R.M. 'Sensory Reception in Bacteria', Symposium of the Society for Experimental Biology, No XXXV. *Prokaryotic and Eukaryotic Flagella*, eds W.B. Amos and J.G. Duckett (Cambridge University Press, 1982) pp. 77–104

Ottow, J.C.G. 'Ecology, Physiology and Genetics of Pili and Fimbriae', *Annual Reviews of Microbiology* (1975), *29*, 79–108

Taylor, B.L. 'Role of the Protonmotive Force in Sensory Transduction in Bacteria', *Annual Reviews of Microbiology* (1983), *37*, 551–73

5 THE CELL ENVELOPE IN BACTERIAL DISEASE

The Role of the Cell Envelope

For an organism to cause disease it must be capable of fulfilling the minimum requirements of any pathogen:

(a) it must first gain entry into the host;
(b) once within the host it must be able to survive and multiply;
(c) it must resist or avoid various host defence mechanisms; and
(d) the host tissues must be damaged in some way.

In many instances the substances involved in the above process are extracellular, such as toxins and enzymes, but in others, structural components of the bacterial envelope are important. Although the anatomy of the envelope of pathogens is highly variable between different organisms, the structures contributing to the ability of an organism to cause disease can be broadly grouped into four categories namely: (a) the capsular layer; (b) lipopolysaccharides and other amphiphiles; (c) protein components; (d) surface appendages.

The Capsular Layer

Capsules and capsular antigens are considered to be outer envelope polymers, usually high-molecular-weight acidic polysaccharides, which surround the bacterial cell in the form of a hydrophilic gel. Polysaccharides adhering weakly to the cell are known as slime layers.

The thickness of capsular gels may vary enormously. In some strains the capsule may be visualised microscopically as a zone 0.2–1.0 μm thick surrounding the cell in negatively stained preparations (Figure 5.1). In others a microcapsule of 10–30 nm thickness is only detectable using sensitive chemical or immunological techniques.

The size and form of capsule produced may also vary with the physiological state of the cell. Growth conditions in artificial laboratory cultures clearly influence capsular polysaccharide production. The effects of a particular nutrient limitation or change in growth rate, in the controlled environment of the chemostat, have shown that exopolysaccharide production by different strains responds in different ways to a given environment. In the same way, an organism entering host tissues encounters complex environmental conditions which may markedly influence capsular polysaccharide production.

147

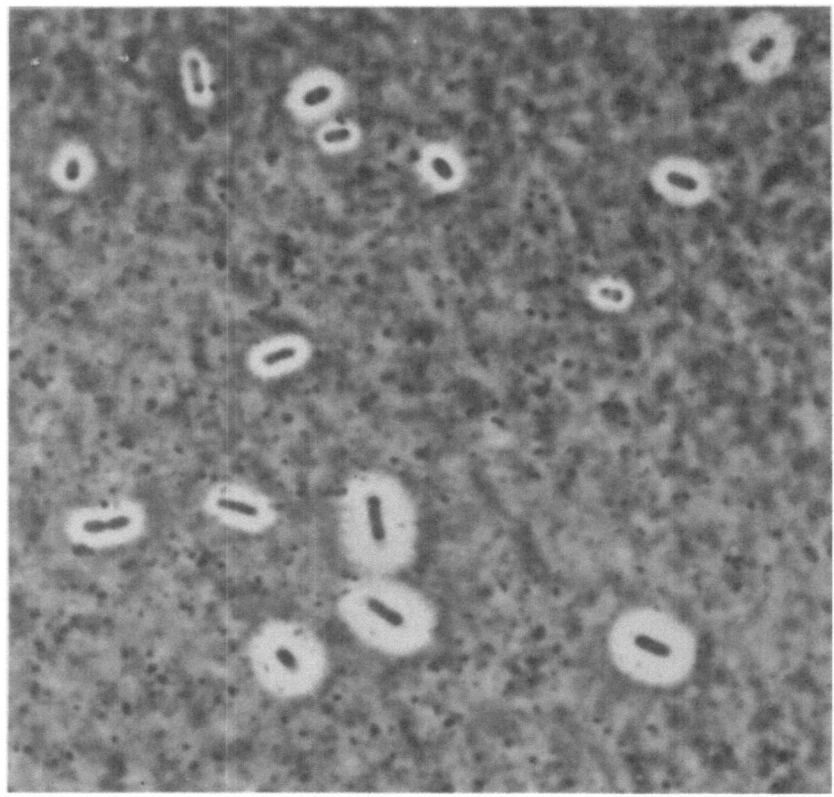

Figure 5.1: Heavily Encapsulated Strain of *Escherichia coli* Viewed by Negative Staining. Note variation in size of capsule. (× 2000).

Conditions *in vivo* are nutritionally more than adequate to support the growth of a wide range of microorganisms. The availability of free iron may be limited, but many pathogens are able to extract iron from the environment by using high-affinity iron acquisition systems. In addition, the pH and redox potential are favourable and a supply of oxygen is available for aerobic growth. However, the limitation of microbial populations *in vivo* is achieved by the potent cellular and humoral antibacterial defence mechanisms. Microbial products resisting these defences, e.g. capsules, are selected *in vivo* because they enhance the survival of the pathogen.

Encapsulation and Virulence

Many pathogens taken directly from body tissues or in primary culture are encapsulated. Pathogens including *Streptococcus pneumoniae*, pyogenic streptococci, *Staphylococcus aureus*, *Klebsiella pneumoniae*, *Neisseria meningitidis*, *Escherichia coli*, *Salmonella typhi* and *Haemophilus influenzae* are invariably encapsulated when examined fresh from an established infection. Some of these organisms lose their capsules on subculture. This may be a phenotypic modification (due to the nature of the artificial laboratory environment) as demonstrated with acapsular strains of *S. aureus*. These may regain their capsules under appropriate conditions. Alternatively it may be a genotypic change (mutation) as seen in spontaneous variants of pneumococci. Such acapsular mutants have been obtained from a variety of different pathogens and have been shown to be of much lower virulence than the parent. Recovery of encapsulated revertants has also been shown by passage of non-capsulated mutants through a suitable host animal, and these invariably possess the virulence expressed by the original parent. The well-known experiment of Griffith in 1928 demonstrated the crucial importance of the capsule to the virulence of pneumococci. When killed, smooth (encapsulated) cells, and live, rough (acapsular) cells were inoculated simultaneously into a mouse, the animal died from infection with virulent organisms encapsulated with the serotype of the smooth parent. This result indicated transformation (rather than reversion) of the avirulent strain to virulence *in vivo*, by acquisition of a capsule. In fact only a few bacteria would have inherited the antigen, but these would be able to multiply while the acapsular organisms were destroyed. It is evident from this type of observation that organisms which are capable of survival in the extracellular tissues are often selected to produce capsular material *in vivo* which becomes redundant in artificial culture.

An alternative method of studying the importance of capsular polysaccharides in the pathogenesis of infection is the removal of the polysac-

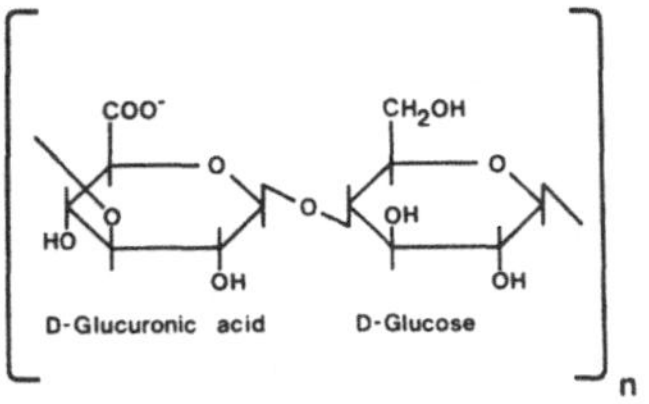

Figure 5.2: Structure of the Disaccharide Repeat Unit of the Type III Capsular Polysaccharide from *Streptococcus pneumoniae*

charide *in vitro* by enzymic digestion. Again the pneumococcus provides an example. The type III capsular antigen (Figure 5.2), a serotype often associated with highly virulent infection, was found to be digested by an enzyme preparation from an organism isolated from soil. When the bacterial cells were decapsulated *in vitro* and administered to mice, the organisms were unable to cause infection. Unlike untreated cells they were rapidly phagocytosed. In an effort to find antimicrobial preparations for use against human infection the enzyme was also used in large doses to protect and cure experimental animal infections. Although some success was achieved, the problems of such treatments outweighed the benefits, but a role for capsular polysaccharide *in vivo* was clearly demonstrated.

The quantity of capsular polysaccharide is apparently of importance in a variety of pathogens. *Klebsiella pneumoniae*, *Salmonella typhi*, *S. typhimurium*, *E. coli*, *Streptococcus pneumoniae* and *Staphylococcus aureus* have all been reported to show a relationship between the amount of capsular material and virulence, resistance to phagocytosis, or resistance to humoral host defences. It has been suggested in the case of *E. coli* that the amount of K-antigen, as measured by its ability to coat red blood cells and inhibit their aggregation by haemagglutinating antiserum, is the most important factor determining resistance to the complement-mediated bactericidal activity of serum. Other studies have found evidence that the type of capsular polysaccharide may be more important.

Bacterial species of medical importance have frequently been subdivided on the basis of their capsular type for the purpose of epidemiological study of infection. These typing schemes, which are usually based upon serological analysis of the capsular antigen, have demonstrated the wide variety of antigenic types produced by different strains of the same species. *S. pneumoniae* may exhibit any one of at least 80 different capsular types. Among the Gram-negatives, *E. coli* has 72 polysaccharide K (Kapsel) antigens currently recognised, although many more undefined capsular types also exist in this species. Within this diversity of structure and chemical composition, a few capsular types are more often associated with infection than others. All pneumococci causing infection are encapsulated, but the type III pneumococcal capsule is associated with a high morbidity. Similarly, 12 of the recognised capsular serotypes account for 75 per cent of all pneumococcal infections. While this may reflect uneven distribution of pneumococcal capsular types in the environment to some extent, distribution does not account for many of the correlations noted between capsular type and virulence. Strains of *E. coli* causing urinary tract infection and other extra-intestinal infections are frequently found to possess one of 6 capsular (K) antigens, namely: K1, K2, K3, K5, K12 and K13. K1 is particularly notable in being found on the surface of approximately 80 per cent of *E. coli* strains from neonatal meningitis — a disproportionately high frequency considering the K1 carriage rate in normal faeces (the source of neonatal infection during birth) is about 12 per cent.

Although information gained from surveys relating antigen type, and source and severity of infection has provided some insight into which capsular polysaccharides are important in disease, it is imperative to take this information in its true context. The ability of an organism to fulfil the minimum requirements of a pathogen (p. 147) depends upon several factors, all of which must be functional. Other non-essential factors may further enhance strain virulence. The comparison of wild strains isolated from infection with respect to serotype alone does not take into account the presence or absence of these unknown variables and hence it is of limited value. It is for this reason that two strains possessing an identical capsular antigen, known to contribute to virulence may differ widely in their ability to cause infection. More precise data on the role of surface structures, including capsules, can be obtained through the comparison of well-characterised mutants, or genetically constructed strains. Another approach is to alter phenotypically the structure of a pathogen by the use of continuous culture to grow the bacteria under various controlled conditions, or by growth in perforated plastic chambers implanted subcutaneously in experimental animals so as to expose the bacteria to the environmental conditions *in vivo*. Bacteria obtained in this way can be examined for alterations in surface structure which may accompany a change in virulence of the strain.

Unfortunately, information on the role of surface components obtained in this way, is relatively scarce. However, when available it serves to give clear insight into the part played by bacterial surface structures in disease.

Antiphagocytic Properties of Capsules

Bacterial strains which possess capsules are almost always more resistant to phagocytosis than unencapsulated strains of the same species. It is not surprising therefore that capsules are found to be important virulence factors in extracellular pathogens which are rapidly destroyed following ingestion by phagocytes. Despite the long-standing recognition of a role for capsular polysaccharides in resistance to phagocytosis, the mechanism of resistance has only recently been clarified.

Phagocytic cells are a fundamental part of the host defences. They are found in all vertebrates and many invertebrates, suggesting an ancient origin. In man the main phagocytic cells are the rapidly mobilised, short living polymorphonuclear leucocytes, mainly neutrophilic granulocytes; and the longer living mononuclear leucocytes and fixed macrophages of the reticulo-endothelial system. The importance of phagocytic cells is revealed by the increased susceptibility to infection of individuals suffering congenital or drug-induced impairment of phagocytic function.

Two levels of phagocytic activity are possible. The process termed 'surface phagocytosis' refers to the ingestion of particles (including microorganisms) which are not coated with opsonins (immunoglobulin G and

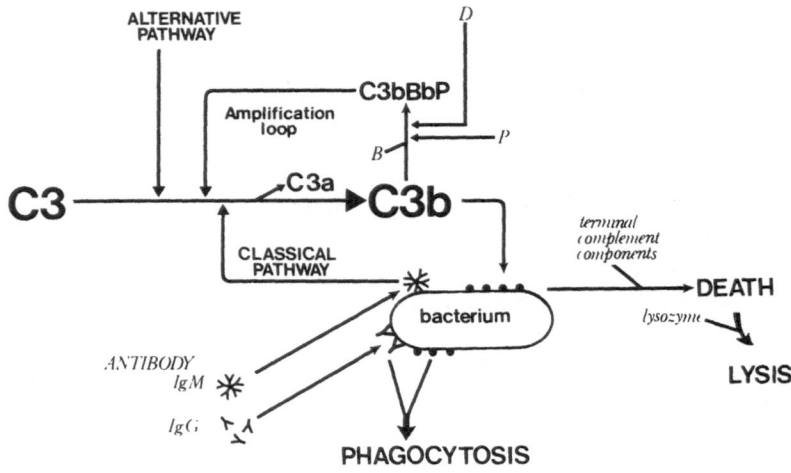

Figure 5.3: The Role of Complement in the Elimination of Microorganisms from the Body. The central reaction generating the opsonically active protein C3b may be initiated by either the classical or alternative pathways. C3b deposited on the bacterial surface enhances phagocytosis, as does the binding of IgG antibody to surface antigens. IgM antibody mediates phagocytosis mainly through classical pathway complement activation. Fixation of complement at the surface of Gram-negative organisms can also lead to death of the cell, effected by the later complement components. The activation of complement is amplified by catalytic feedback of the alternative pathway. Three factors; properdin (P), B and D interact with C3b to form the alternative pathway C3 convertse (C3b BbP).

the activated complement component C3b, Figure 5.3). This is a relatively slow and inefficient system compared with the vigorous opsonin-mediated phagocytosis. The physicochemical nature of the surface of a particle is an important determinant of adherence to, and subsequent ingestion by, the phagocyte during surface phagocytosis. Some surfaces are therefore inherently more resistant to phagocytosis than others.

Recent work has demonstrated the importance of the degree of hydrophobicity of a particle surface in its susceptibility to surface phagocytosis. The more hydrophilic was the bacterial surface, as measured by the contact angle made between a saline solution and dried films of bacteria, the more resistant was the organism to phagocytosis by neutrophils in the absence of opsonins. This expounds the idea of the hydrophilic and negatively charged nature of polysaccharide capsules retarding surface phagocytosis.

The view of opsonisation widely held until recently was that it serves to mask and thereby overcome the ability of a particle to avoid contact or ingestion by the phagocyte, i.e. the opsonised surface becomes physicochemically more favourable to phagocytosis. This may well be a sound hypothesis, but in view of the observation that very small quantities of opsonins can render a heavily encapsulated organism susceptible to phagocytosis, this passive role appears to be of secondary importance.

Current emphasis is now given to the opsonin as a specific ligand whose conformation is recognised by receptors located on the plasma membrane of the leucocyte. Interaction of C3b or the Fc region of IgG with its receptor constitutes immune adherence of a particle and stimulates ingestion of that particle with subsequent degranulation (release of lysozomal enzymes into the phagosome). This is associated with an increase in the rate of oxygen consumption by the leucocytes, the so-called respiratory burst. Antiphagocytic capsules appear to act by interfering with the recognition of the bacterium via the leucocyte receptors. In order to understand how the capsule achieves this, we must first examine the operation of the system during the elimination of a non-pathogenic microbe.

A non-encapsulated strain of *S. pneumoniae* may penetrate the tissues, for instance after accidental mucous membrane damage. The natural (non-immune) antibody population includes antibodies specific for commonly encountered bacterial substances such as peptidoglycan, teichoic acids and enterobacterial common antigen. This system allows the body to recognise bacteria as foreign, and phagocytic cells will ingest and destroy the opsonised invader. However, another recognition system is able to react without previous experience of those foreign substances, and it therefore operates in the non-primed host. The complement system activated via the so called alternative pathway provides the effector for this hitherto undefined recognition mechanism (Figure 5.4). Certain substances such as the yeast polysaccharide zymosan, peptidoglycan and bacterial lipopolysaccharide initiate activation of the alternative pathway independently of antibody. Activation occurs because these substances provide an 'activating surface' which protects spontaneously generated C3b from decomposition. The factors governing the suitability of a surface in preventing C3b decomposition (and *ipso facto* recognition of the surface as foreign) resulting in activation of the alternative complement pathway, are not yet understood.

Contrary to the implication in its naming, the alternative pathway is the more primitive mechanism of generating an inflammatory response — the antibody dependent (classical) pathway having evolved later as a specific adaptor system.

Whichever mechanism is involved in the recognition of the bacterium as foreign, complement activation will follow, resulting in the deposition of C3b on the surface. When specific IgG is available to bind to the organism, the Fc regions will be exposed on the surface, and recognition of these ligands stimulates the phagocytes to ingest and destroy the bacterium.

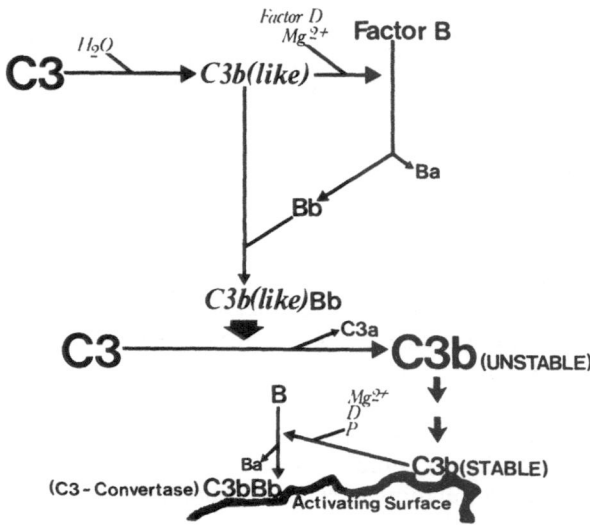

Figure 5.4: The Alternative Pathway of Complement Activation. C3 is spontaneously hydrolysed without cleavage of C3a, to become C3b (like). This has the ability to cleave factor B. Combination of C3b (like) with Bb forms a C3 convertase capable of cleaving further C3 into C3b and the anaphylotoxin C3a. Fluid phase C3b is unstable and rapidly inactivated by control proteins, but the availability of a suitable protective surface serves to bind and stabilise the C3b. Stabilisation results in activation of the alternative pathway and constitutes recognition of the protective surface as foreign. C3b can then cleave more factor B to form the true C3 convertase C3bBbP, itself stabilised by the enhancing control protein Properdin (P).

Capsules of Gram-positive Bacteria. The role of the capsule in resisting phagocytosis has been studied in a number of organisms. Studies with the heavily encapsulated strain *Staphylococcus aureus* M have shown that the cell wall peptidoglycan will fix complement by either the classical or alternative pathways, but in encapsulated strains the C3b becomes bound beneath the capsule at a depth which is inaccessible to the phagocyte receptors (Figure 5.5). The capsule in this case is interfering with the exposure of opsonins at the external surface of the organism. For this mechanism to be effective the capsular polysaccharide must itself be incapable of acting as an activating surface in the absence of specific anti-capsular antibody. This property appears to be a common feature of many capsular polysaccharides, but one which is variable with composition and structure, and hence capsular type.

Most pneumococcal capsules do not activate the alternative pathway, and they also do not seem to interfere with access to the cell wall where

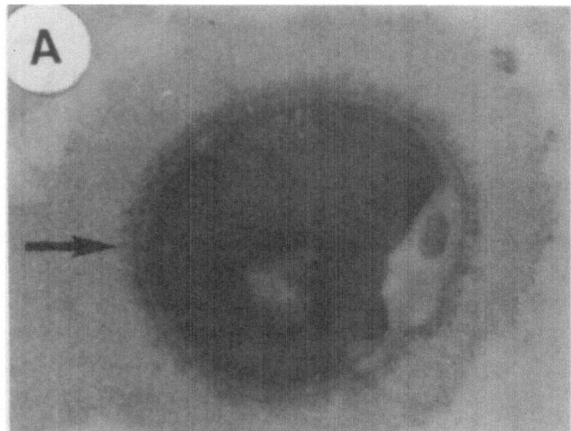

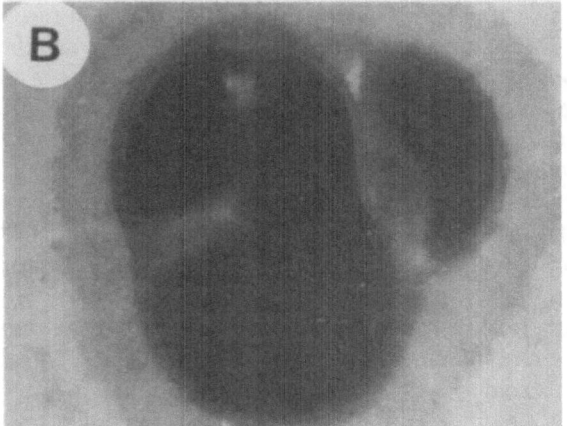

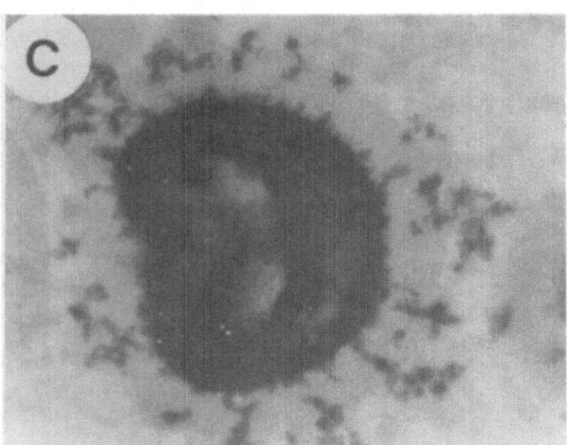

Figure 5.5: Electron Micrographs of *Staphylococcus aureus* M Showing the Disposition of C3 Fixed on the Bacterial Surface. *S. aureus* incubated in normal serum, and then treated with peroxidase labelled anti-C3 (A), or buffer as control (B). C3 is visible as a densely staining region at the cell wall (arrow), beneath the external surface. When immune antibody is present in the serum (C), the C3 is deposited throughout the capsular layer including the true external surface where it is accessible to leucocyte receptors. Reproduced by permission, Verbrugh H.A., Peterson P.K., Nguyen B.T., Sisson S.P. and Kim Y., *J. Immunology* (1982) *129*, 1681.

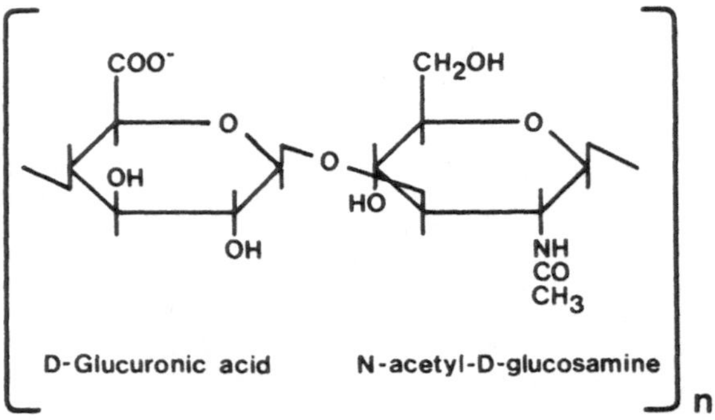

Figure 5.6: Structure of the Disaccharide Repeat Unit of Hyaluronic Acid: the Capsular Polysaccharide of Group A Haemolytic Streptococci

fixation of C3b can take place. Complement factors which have penetrated the capsule can be demonstrated using immunoelectron microscopy, and are found to be associated with deeper envelope components, again out of reach of the leucocytes. When specific anticapsular antibody is available, either from previous exposure or formed *in vivo* during the course of an infection, it is protective to the host. Immunoglobulin will bind to the capsular antigen and, in the case of IgG antibody, the Fc portion alone is opsonic. Although bound IgM antibody is itself not a good opsonin it is very efficient at activating complement via the classical pathway, generating the leucocyte chemotaxins C3a and C5a and the major opsonin C3b bound to the bacterial surface where interaction with phagocytic receptors can take place.

An added complication occurs when the surface layer is also a poor immunogen. Group A haemolytic streptococci produce a capsule of hyaluronic acid (D-glucuronic acid $\beta 1 \rightarrow 3$ *N*-acetyl-D-glucosamine, in $\beta 1 \rightarrow 4$ linkage, Figure 5.6). This polymer is identical to the extracellular hyaluronic acid of vertebrate connective tissue and synovial fluid and hence will not elicit an immune response as the capsule is recognised as 'self'. For the same reason the polysaccharide will not normally activate complement by the alternative pathway since reaction to the same immunodeterminants throughout the body would result in uncontrolled systemic inflammation. Further examples of pathogens which avoid host defences by sharing host antigens will be discussed later.

The molecular basis for resistance to complement activation has been studied in the group B streptococci. The type III capsule of this organism has been shown to inhibit activation of the alternative complement pathway. The polymer consists of repeating pentasaccharide units each with a terminal sialic acid residue (p. 157) which masks the underlying galactopyranose moiety. Cleavage of terminal sialic acid residues by neuraminidase exposed the underlying sugar residues revealing a surface capable of activating complement by the alternative pathway. Further investigations have demonstrated that it is the tertiary molecular conformation of this polysaccharide that is of prime importance in resistance to complement activation, and not the presence of the terminal sialic acid residue alone.

Another Gram-positive organism possessing capsular antigens conferring virulence is *Bacillus anthracis.* Only one capsular type is known: a polymer of D-glutamic acid. This polymer is thought to confer resistance to phagocytosis by inhibiting the absorption of serum opsonins. It is also implicated, together with the anthrax toxin, in interference with horse serum bactericidins including β-lysins, and basic polypeptides. The capsule of *B. anthracis* is thought to be important only in the establishment of infection, after which the major aggressive factor is the three-component toxin.

Capsules of Gram-negative Bacteria. Unlike those of Gram-positive bacteria, the capsular polysaccharides of Gram-negative bacteria, of which the best studied are the K-antigens of Enterobacteriaceae, appear to restrict the access of opsonins to deeper structures. Impeding IgG and complement proteins from penetrating the capsular layer may be important in preventing the lipopolysaccharide and other complement activating substances from initiating the complement sequence in close proximity to the outer membrane. This is significant because many Gram-negative organisms are susceptible to the direct bactericidal action of the terminal complement sequence which is capable of inflicting irreparable membrane damage upon the cell.

To be effective in preventing or reducing susceptibility to phagocytosis these polysaccharides must share with those of Gram-positive organisms the property of failing to fix complement in the absence of specific antibody. This many of them appear to do, and additionally they may be of low immunogenicity. A further feature of these antigens is the great diversity of antigenic types displayed at the surface. This is an important means by which the likelihood of previous host experience with the surface antigens is reduced; another example of which is found in the rhinoviruses which are responsible for the common cold.

The capsular antigens of enterobacteria have been examined serologically and chemically since their description as substances which prevented the agglutination of live cells by O-specific antisera. The K1 capsular antigen as previously noted deserves special mention for its association with invasive *E. coli* infection. It is a homopolymer of $\alpha 2 \rightarrow 8$ linked *N*-acetylneuraminic acid (Figure 5.7). *N*-acyl derivatives of neuraminic acid are collectively known as sialic acids, and are important constituents of mammalian glycoproteins and glycolipids located in the cell coat. *N*-acetylneuraminic acid is found in human tissue, and the presence of this polysaccharide on the surface of bacteria represents another example of avoidance of host defences by exhibiting the same antigenic determinants as host tissue.

The K1 antigen has been found to undergo phase variation to an *O*-acetylated form (Figure 5.7). The *O*-acetylated derivative of the polymer appears to be more immunogenic than the non-acetylated form, but otherwise no differences have been reported in relation to avoidance of host defences, including ability to activate the alternative complement pathway, between the two forms.

The structure of several other K-antigens have been reported by Jann and co-workers. Of these K5 deserves mention as a further example of a non-immunogenic K-antigen. The K5 polysaccharide composed of repeating disaccharide units of $\alpha 1 \rightarrow 4$-linked glucuronyl $\beta 1 \rightarrow 4$ *N*-acetylglucosamine closely resembles desulpho-heparin, a precursor of the natural blood anticoagulant heparin.

Neisseria meningitidis, the organism responsible for epidemic purulent meningitis is subdivided on the basis of its capsular antigen. Nine serogroups are currently recognised: A, B, C, D, X, Y, Z, W135 and 29E. Serogroups A, B and C are responsible for most of the meningitis outbreaks reported. The group B polysaccharide is an identical polymer to the K1 antigen of *E. coli*, while the structure of the group C antigen is similar, being $\alpha 2 \rightarrow 9$ linked *N*-acetylneuraminic acid, *O*-acetylated at C-7 and C-8. Studies on the group B meningococcus have shown that loss of the capsular polysaccharide biosynthetic capability in isogenic mutants of a virulent strain resulted in loss of virulence for mice. This virulence was regained upon reversion to the wild-type phenotype. Like the K1 antigen, the group B capsular polysaccharide has low immunogenicity in humans, but groups A and C have been shown to be effective vaccines. Little information concerning the role of these capsular polysaccharides is available, and while they are likely to be indispensible for survival *in vivo*, other factors, probably surface located, are likely to be important in the pathogenicity of meningococcal meningitis.

The capsular polysaccharides of other Gram-negative pathogens are also associated with decreased susceptibility to phagocytosis. The Vi antigen of *Salmonella typhi*, *S. paratyphi* and a few strains of *E. coli* is a homopolymer of *N*-acetyl-D-galactosamine uronic acid. While being of some importance in discouraging phagocytosis, it is not essential for *Salmonella* infection since many species causing disease do not display any capsular layer. They are facultative intracellular parasites, being able to avoid the bactericidal systems within phagocytic cells to some extent. *Neisseria gonorrhoea* is similarly able to resist immediate killing by polymorphonuclear leucocytes, and it is often found within these cells in urethral pus. However, fluorescence data suggest that capsular structures are present on freshly isolated organisms and they appear to be involved in resistance to

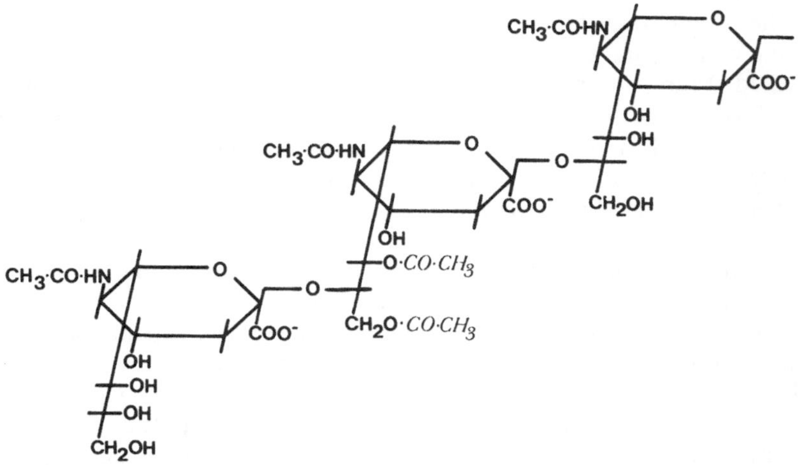

Figure 5.7: Structure of the K1 Antigen of *E. coli*. The central *N*-acetylneuraminic acid residue is shown *O*-acetylated at C-7 and C-9 as is found in form variants of this polysaccharide.

non-antibody mediated phagocytosis — presumably 'surface phagocytosis'. Unfortunately these capsular substances are little known since they are so rapidly lost from the organisms during subculture in artificial media.

Another facultative intracellular parasite, *Yersinia pestis* produces a capsular layer of protein, lipid and polysaccharide termed fraction 1. In scanning electron micrographs it appears as a granular layer covering the bacterial surface. It confers resistance to phagocytosis, and was once thought to be critical in the pathogenesis of plague. However, fraction 1-negative mutants have only partially reduced virulence, as might be expected from a pathogen otherwise capable of intracellular survival.

The obligate intracellular pathogen *Brucella abortus* has only recently been confirmed as capsule producing. Like *N. gonorrhoeae* this organism does not synthesise capsular material outside the body. In addition the capsule is easily removed from the cell by washing, and only visible after stabilisation with anticapsular antibody and staining for electron microscopy with the cationic dye ruthenium red which has an affinity for strongly anionic polymers such as polysaccharides.

Capsules in Resistance to Serum Complement

As noted above many Gram-negative bacteria are sensitive to the direct killing action of serum complement. This bactericidal activity depends upon the later members of the complement sequence, and is thought to be important as a host defence against invasion by Gram-negative bacteria. When activated, the terminal complement components C5 to C9 form the so called membrane attack complex (MAC), a large ring-form complex capable of inflicting irreparable membrane damage. When complement is activated at the outer membrane a series of events ensues which culminates in the formation of a channel through both the inner and outer membranes, perhaps located at the adhesion points (p. 82). The channel appears to allow free passage of low and high-molecular weight cytoplasmic constituents across the membrane. This leakage dissipates the membrane potential derived from the proton gradient formed across the inner membrane. The result is death of the bacterium, and together with the action of lysozyme the reaction often causes cell lysis (Figure 5.3).

In contrast the terminal complement sequence is not capable of directly killing Gram-positive organisms, probably because of the restricted access to the cytoplasmic membrane conferred by the thick peptidoglycan wall.

Less is known about the involvement of capsular polysaccharides in protection against the complement-mediated bactericidal activity than resistance to phagocytic ingestion. *E. coli* K antigens are known to restrict the access of complement proteins to the deeper layers of the bacterial envelope which are capable of initiating the complement sequence. However, many K-positive strains are sensitive to complement killing implying that the barrier is by no means always effective. The sensitivity of strains depends upon a variety of factors, some known and some merely postulated. *In vitro* studies have shown the thickness of capsular material, concentration of serum and K serotype to be important. Numerous studies have been conducted to compare the serum susceptibility of serologically classified strains of Gram-negative pathogens. No conclusive findings have resulted from these since unknown factors, perhaps working in concert with the capsular antigen, control this behaviour. The K1 antigen, thought to be a poor activator of the alternative pathway has been clearly shown to confer resistance in strains able to grow in serum. K1-negative mutants were rapidly killed by incubation in serum. However, when other *E. coli* were genetically constructed so that they inherited the identical capsular K1 polymer, both by introduction of a cloned fragment encoding K1 biosynthesis and also by conjugation, the outcome was variable. While some strains were protected to some degree, others gained no detectable advantage in their ability to survive in serum. Studies investigating the K27 antigen of *E. coli*, a demonstrably thick capsule, have failed to find any role for this polysaccharide in complement resistance. In contrast, a strain of *Citrobacter* carrying the Vi antigen was clearly more resistant to complement killing when grown at 37°C, than 42°C. The Vi antigen is only produced at the lower temperature indicating that the Vi polysaccharide is protective to the cell.

Slime Polysaccharides

Surface Polysaccharides in Cariogenicity. The production of dental caries (tooth decay) is strongly associated with the presence of *Streptococcus mutans.* Other oral streptococci, e.g. *S. sanguis* and *S. salivarius,* and lactobacilli are also found in the lesions, but are not considered to be primary pathogens, except in caries initiated in dental fissures and pits where food has become impacted and strong adhesive properties are not essential.

A major virulence factor of *S. mutans* is the production of extracellular polysaccharides, glucans and fructans, which are responsible for adhesion of the bacteria to the smooth surface of teeth. In addition the organisms excrete organic acids, which may be concentrated locally, to which *S. mutans* is relatively tolerant. These organic acids are responsible for the destruction of dental tissue. Experiments have shown that precoating of the tooth surface, or *S. mutans* cells, with glucan allows adherence of the bacteria to the tooth but the bacteria stick far more firmly when glucan synthesis takes place in the presence of sucrose *in situ* on the tooth surface.

The surface enamel of teeth consists mainly of hydroxyapatite, a form of crystalline calcium phosphate. Glycoproteins from saliva bind strongly to the Ca^{2+} ions in the crystal lattice and form a very thin surface film called the enamel pellicle. Bacterial surfaces associate with the pellicle and become irreversibly bound by polysaccharides. A matrix builds up consisting of bacteria, polysaccharide and oral cell debris which is commonly known as dental plaque. The cariogenesis of *S. mutans* is clearly related to the formation of plaque from bacterial glucans and this in turn is dependent upon the presence of dietary sugars, particularly sucrose. *S. mutans* is able to form glucans by constitutive production of the extracellular enzyme glucosyl transferase (GTase). This reaction:

$$\text{Sucrose} \xrightarrow{\text{GTase}} \text{Glucan polymer} + \text{fructose}$$

produces a polymer which consists of $\alpha 1 \rightarrow 6$ linked glucose residues branched with a variable proportion of $\alpha 1 \rightarrow 3$ linkages. The proportion of $\alpha 1 \rightarrow 3$ linkages in the polysaccharides determines the degree to which the polymer is soluble, a high proportion decreasing the solubility. Some $\alpha 1 \rightarrow 2$ and $\alpha 1 \rightarrow 4$ linkages are also found. Fructans are formed less frequently. They are thought to have a $\beta 2 \rightarrow 1$ fructo-furanoside linkage.

Experiments with gnotobiotic animals, monoinfected with *S. mutans* have shown the organism to be responsible for severe caries. Glucosyl transferase-deficient mutants however produce no insoluble glucans, and this has been found to significantly reduce cariogenic potential in experimental animals. Another class of mutants has been found which produces a greater proportion of water-soluble glucans. The deficiency may therefore lie in the $\alpha 1\rightarrow 3$ glucosidic linking activity of the glucosyl transferase. These mutants also are unable to form plaque and caries on smooth teeth surfaces. It seems that in addition to the powerful adhesive effect, insoluble glucans also produce a barrier to the diffusion of excreted lactic acid generated by metabolic activity. This localisation of the acid appears to be critical in the demineralisation of the teeth, and its greater acid tolerance may distinguish *S. mutans* from other oral steptococci as the sole true cariogenic species.

Other experiments have shown the dependence of glucan synthesis and cariogenesis upon the availability of sucrose. When *S. mutans* is grown in glucose instead of sucrose it is unable to produce glucans (Figure 5.8), and experimental animals fed glucose instead of sucrose have a marked reduction in plaque formation and carious lesions. The dependence of carie formation upon the synthesis of insoluble glucan from sucrose has prompted studies to examine various preparations of glucan hydrolysing enzymes for their anti-plaque and anti-cariogenic potential. The studies have produced some conflicting results, but in general the findings have shown that $\alpha 1\rightarrow 6$ glucanases (dextranases), e.g. from *Penicillium funiculosum*, have a limited capacity to degrade extracellular glucans of *S. mutans in vitro*. Studies with this and other $\alpha 1\rightarrow 6$ glucanases in human trials have indicated variable results while those using animal models have often shown success in preventing plaque formation. An $\alpha 1\rightarrow 3$ glucanase (mutanase) purified from another fungus *Trichoderma harzianum* has been shown to reduce plaque formation and colonisation of plaque by *S. mutans*.

The Slime Polysaccharide of Pseudomonas. Pseudomonas aeruginosa isolated from chronic respiratory tract infections of individuals with cystic fibrosis (CF) is very often of a distinctive mucoid morphology. Approximately 80 per cent of all *P. aeruginosa* strains taken from CF patients are mucoid, while only 1–2 per cent of clinical isolates from the general population show this distinctive feature. The mucoid appearance is due to a genetic alteration, perhaps a mutation influencing a bacterial control mechanism, resulting in the production of copious amounts of alginate. This *O*-acetylated exopolysaccharide is composed of $1\rightarrow 4$ linked D-mannuronic acid and variable proportions of its 5-epimer L-guluronic acid. Alginate production is unstable and non-mucoid revertants are found amongst the mucoid colonies when cultured *in vitro*.

When first established in the lung of CF patients, *P. aeruginosa* is of the non-mucoid form. Alginate producing organisms gradually come to predominate, and the chronic persistance of these forms, despite intensive physiotherapy and antibiotic therapy, suggests that a clear survival advantage is gained by the organism from alginate synthesis.

In vitro studies have suggested that the mucoid coating may protect the

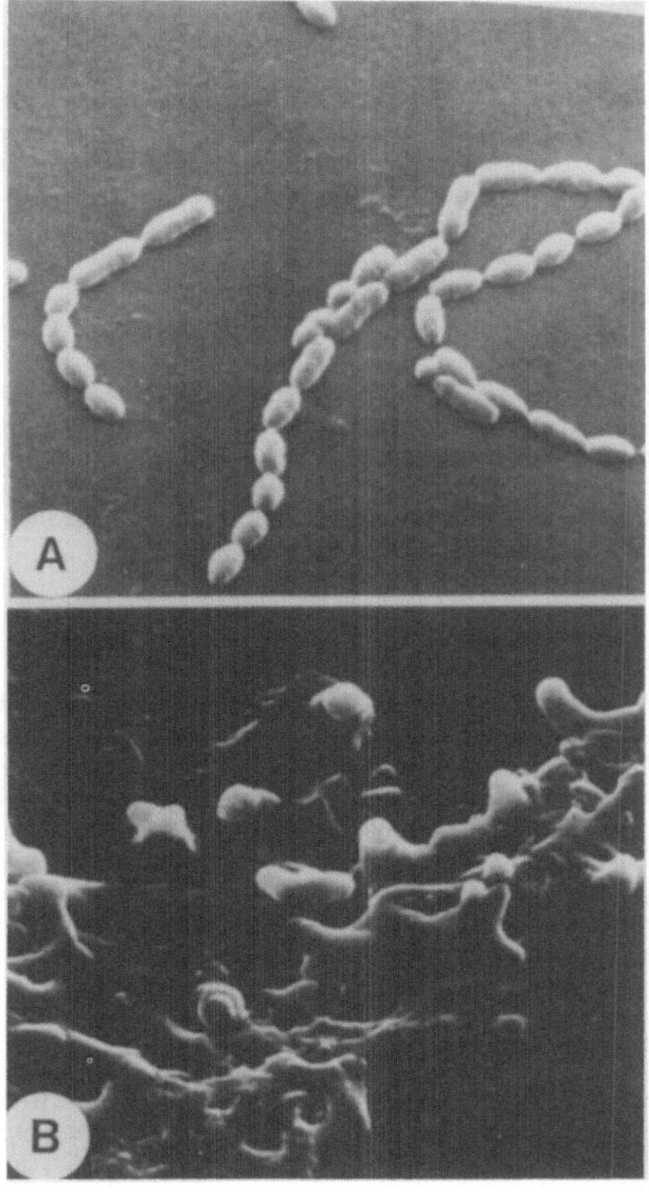

Figure 5.8: Scanning Electron Micrograph of *Streptococcus mutans* grown (A) in Broth, and (B) in the Same Broth Containing 5 per cent Sucrose. The abundance of sucrose-dependent slime polysaccharide adhering the bacteria to the smooth surface is clearly visible. Reproduced by permission: Hamada, S., Mizuno, J., Murayama, Y., Ooshima, T., Masuda, N..and Sobue, S. (1976) *Infect. Immunity.*, *12*, 1415.

bacteria from phagocytic cells, antibodies and antibiotics. Comparison of mucoid and non-mucoid strains *in vitro* have indicated that the alginate may be antiphagocytic by blocking or masking the immunodeterminants for opsonic antibody. However, a study utilising a guinea-pig model of experimental pneumonia to examine the effect of alginate production on clearance and killing by intrapulmonary phagocytic cells failed to find any differences between mucoid strains and their isogenic non-mucoid revertants. This finding suggests that the alginate does not impede mechanisms of intrapulmonary killing within the guinea-pig lung, and the factors selecting for mucoid strains in CF patients, together with the manner of survival advantage gained by alginate production, remain unclear.

Slime Polysaccharide in Staphylococcus epidermidis. This organism, a common commensal of the skin, has become an important opportunistic pathogen associated with medical devices such as indwelling catheters and shunts. Staphylococci isolated from these sources often produce a polysaccharide slime which is thought to be important in the adherence of the organism to the smooth surface of the foreign body.

Lipopolysaccharides and Other Bacterial Amphiphiles

Amphiphiles are a diverse range of polymers that share the common features of possessing both hydrophobic and hydrophilic regions. A variety of amphiphiles are found in bacterial envelopes including lipopolysaccharides, enterobacterial common antigen, lipoprotein, lipoteichoic acid and others. Some of these molecules have a role in the pathogenesis of bacterial disease. By far the most widely studied molecules of this type are the lipopolysaccharides, found exclusively in the outer membrane of Gram-negative bacteria. Our knowledge of other bacterial amphiphiles is comparatively poor, partly due to their relatively recent discovery and also because of difficulties in handling and purifying them.

Lipopolysaccharides (LPS)

Much of the current knowledge concerning LPS has come from members of the family Enterobacteriaceae, particularly *Salmonella.* For this reason many of our ideas on the role of LPS in the disease process are related to the virulence of these organisms. However, the basic similarity of structure and biological activity of LPS from many different sources suggest that these ideas may be broadly applicable to the functions of LPS in other pathogens.

The lipopolysaccharides have three distinct regions which are joined by covalent linkages. The hydrophobic portion, termed lipid A, is the part of the molecule known to be responsible for the toxic properties and potent biological effects associated with LPS. The hydrophilic region consists of the O-specific repeating oligosaccharide extending outward from the cell into the environment, and the core oligosaccharide which links the O-side chain to the lipid A. This hydrophilic portion of the LPS molecule is known to be very important in the host-parasite relationship having a role in resisting the innate host defences during infection, and being responsible for the diversity of O-antigenic types on the bacterial surface which can play a role in avoidance of the immune response.

Lipid A. Unlike many Gram-positive pathogens, Gram-negative pathogens do not generally elaborate exotoxins and extracellular enzymes. In many cases the sole mediator of tissue damage in Gram-negative infection is endotoxin, a term which refers to the toxic properties of LPS which reside in the lipid A moiety. It is the only established toxin in infections caused by salmonellae, brucellae, meningococci, gonnococci, *Haemophilus influenzae, Klebsiella pneumoniae* and non-enteric infections due to *E. coli.* That the toxicity of LPS lies in the hydrophobic lipid A has been clearly established, however lipid A alone is insoluble and cannot express biological activity. When the sugars of the inner core are present, i.e. 3-deoxy-D-*manno*-octulosonic acid (KDO), as in Re mutants of *Salmonella,* or when a soluble carrier such as bovine serum albumin (BSA) is attached to the lipid A, it can be dispersed in aqueous solution and possesses full biological activity. The biological effects of LPS are diverse and have been well documented including such phenomena as:

(a) The production of fever in experimental animals, e.g. rabbits (pyrogenicity).
(b) Circulatory disturbances and vascular hyper-reactivity to adrenergic drugs, e.g. the depression of blood pressure in dogs.
(c) Leucopaenia followed by leucocytosis, i.e. rapid depletion of the leucocyte population in the blood followed by an increase in numbers beyond the normal levels.
(d) Schwartzman phenomenon, i.e. haemorrhage at a skin site previously sensitised by injection of a few micrograms of endotoxin, when a similar quantity is administered intravenously 24 hours later.
(e) Non-specific stimulation of β-lymphocytes to undergo blast transformation and proliferation (mitogenicity).
(f) Lethal toxicity, i.e. when sufficient endotoxin is administered by injection irreversible shock, often with diarrhoea, occurs within one to two hours and death ensues.
(g) Non-specific tolerance to endotoxin, i.e. animals given repeated injections of LPS show progressively less response to a given dose.

It is interesting that in addition to reducing pyrogenicity and the other biological effects, tolerance also inhibits the body's response to radiation damage and traumatic shock. The responses to this kind of tissue damage are thought to be due to the effects of released endogenous cellular constituents. Evidence now suggests that the pyrogenicity and certain other biological activities of endotoxin are also caused by these substances, and the endotoxin acts indirectly by somehow mediating the release of physiologically active cell components into the extracellular spaces.

Lipopolysaccharides are thought to bind to eukaryotic cells via the lipid A portion of the molecule. LPS and other bacterial amphiphiles are consistently able to bind to the membrane of erythrocytes. The receptor present on human erythrocytes has been shown to be a lipoglycoprotein, the lipid A binding to the hydrophobic residues of the protein. The binding of LPS to leucocytes, however, appears to be the result of penetration of the lipid A into the phospholipid membrane of the cell. Artificial membranes are known to become less stable when LPS is incorporated into one monolayer. This disruption may be responsible for the leakage of intracellular constituents from eukaryotic cells *in vivo*. It has also been shown that artificial liposomes containing LPS are susceptible to complement-mediated damage when anti-LPS antibodies are present. It should be remembered that many lipopolysaccharides activate complement by the alternative pathway, and the possibility exists that membrane damage may occur by action of the terminal complement components when LPS associates with eukaryotic cells. Another facet of the toxicity may derive from the anaphylatoxins produced by the interaction of LPS and complement. The protein fragments C5a and C3a are generated during the complement activation sequence, and have the capacity to degranulate mast cells. The resulting release of histamine and histamine-like substances from the mast cells can be toxic to host tissues, and in large enough amounts cause a syndrome resembling anaphylaxis. The presence of complement activating substances, including LPS, in the tissues may cause persistent release of these pharmacologically active substances, thereby contributing to the damage inflicted upon the host. However, this damage is not responsible for other endotoxic properties of LPS. When converted into a disaggregated form (e.g. triethylamine salt) LPS does not react with complement, but other endotoxic properties remain unaffected.

LPS with endotoxic properties is not confined to pathogens. Whole cells or cell extracts from commensal organisms often show the same degree of biological activity as similar extracts or killed whole cells from a pathogenic strain. By contrast the potency of lipid A varies with source. Despite the severity of infections caused by them, brucellae and yersiniae possess LPS of relatively low toxicity. Similarly *Pseudomonas aeruginosa* lipid A is less toxic than that derived from Enterobacteriaceae. This is thought to be a result of the predominance of 12 carbon 3-hydroxy fatty acid substituents in the lipid A of pseudomonads, and 14 carbon 3-hydroxy fatty acids in the Enterobacteriaceae, i.e. β-OH myristic acid is replaced by β-OH lauric acid. It has been suggested that the longer fatty acids enhance the disruptive effects of the lipid A on eukaryote membranes. Studies with artificial membranes have also shown that the LPS will penetrate phosphatidylethanolamine bilayers, but not those composed of phosphatidylcholine. The preferential interaction of the lipid A with some phospholipids may partly explain the tissue specificity displayed by endotoxin activity.

Although Gram-negative pathogens possess LPS, and the lipid A of isolated LPS almost invariably shows toxicity when injected into animals, the participation of endotoxin in the pathophysiology of infection remains less clear than might be expected. For instance, difficulties have been encountered in attempts to correlate the endotoxin content of Gram-negative bacteria with pathological effects. The lethal dose of isolated endotoxin for a mouse is equivalent to about 10^9-10^{10} organisms per animal. Mice dying of salmonellosis were in fact found to contain about this number of organisms. However, when mice which had been rendered more sensitive to endotoxin (1000-fold) by *Mycobacterium bovis* (BCG) infection, the dead animals still contained this number indicating that death was not simply due to a lethal burden of endotoxin. Mutants incapable of synthesising lipid A would provide useful information about the role of endotoxin LPS in pathogenicity. Unfortunately, such mutants are not available for study *in vivo*, as the loss of this component is lethal to the cell. As the endotoxic lipid A is a structural component of the cell envelope, interest has been raised about its availability to exert biological activity during the course of Gram-negative infection. It is not clear whether release of endotoxin from the bacterium is necessary, and some evidence suggests that biological activity is exerted by lipid A present in the bacterial envelope. In agreement with this finding is the observation that ampicillin treatment of experimental infections is successful, when it might be expected to release endotoxic material during disruption of the bacterial cells. If release of lipid A is prerequisite for activity, then the endotoxic material would overwhelm the host during therapy. A phenomenon encountered with some Gram-negative organisms growing *in vitro* is budding of the outer-membrane. Blebs containing LPS are released into the growth medium (Figure 5.9). It is not known whether this occurs during growth *in vivo*. Neither is it known whether bacteria grown *in vivo* possess LPS with the same toxic activity. What is clear is that carefully conducted clinical studies have failed to confirm a role for the endotoxic properties of LPS in the manifestations of Gram-negative infections such as typhoid. Although this casts doubt upon its absolute importance in causing damage to host tissues during infection, such a biologically active substance is unlikely to be a mere spectator during infection. Hence, a clearer understanding of its interaction with host tissue will help to unravel

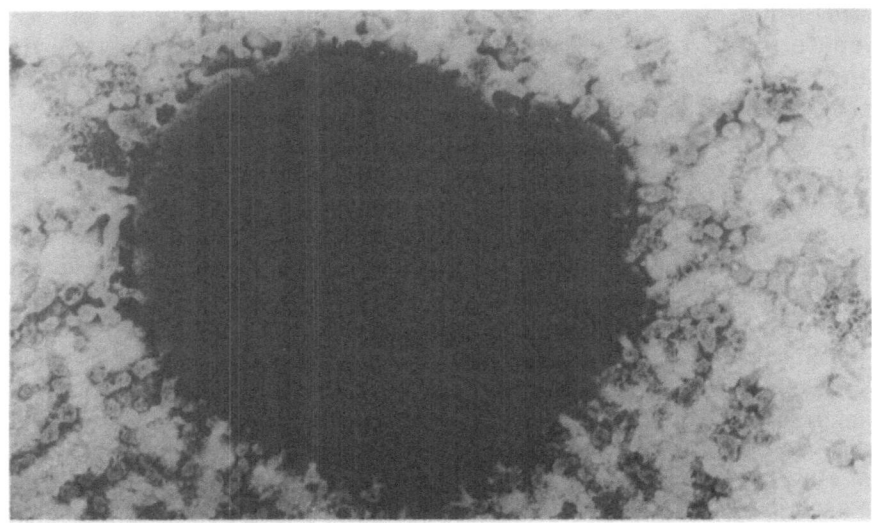

Figure 5.9: Release of Outer Membrane Blebs from the Surface of *Neisseria meningitidis.*
Reproduced by permission: DeVoe, I.W. and Gilchrist, J.E. (1973), *J. Exp. Medicine.*, *138*, 1156.

the true role of lipid A in pathogenicity.

Polysaccharide moiety. The polysaccharide portion of lipopolysaccharide is known to be important in the virulence of many bacteria. Strains lacking polysaccharide side chains (rough strains) are invariably less virulent than those containing such structures.

Within a group of organisms occupying the same ecological niche e.g. *E. coli* or *Salmonella spp.*, an individual strain may display one of a large variety of O antigens. Simply by nature of this diversity, the likelihood of previous immune experience of a given serotype is reduced. Individual members of the group will not be recognised by the immune system of a host as efficiently as organisms displaying a common surface antigen, e.g. peptidoglycan. Thus these complex polysaccharides are instrumental in avoiding the acquired host defences. Antigenic drift of O antigen specificity has been observed in some cases of long-term Gram-negative infection. One antigenic type may gradually alter its specificity to another under the selective pressure of the host's immune response. Evolutionary pressure therefore tends towards increased diversity and this may partly account for the great variety found in these and other surface antigenic structures.

The association between LPS polysaccharide structure and virulence is particularly strong in *Salmonella* species. The genus is a closely related group except for serological differences; highly pathogenic species falling a small number of O groups. *Salmonella* infections are broadly divided into three categories. Those termed enteric fever involve infection of the reticuloendothelial system and bacteraemia. The organisms concerned are capable of survival within the blood and within phagocytic cells. A second category involves localised disseminated infections without bacteraemia, and the third in which the infection is restricted to the intestinal wall is known as gastroenteritis. The ability to cause these infections depends upon penetration of the epithelium of the intestinal wall by the organism, and then survival and multiplication within phagocytes. Studies with *S. typhimurium*, which is particularly virulent for mice, have clarified the penetration process, however the bacterial substances responsible for penetration are unknown. The host specificity expressed by different species of *Salmonella* appears to depend upon intracellular survival rather than ability to penetrate the intestinal wall. Therefore although both *S. typhi* and *S. typhimurium* were found capable of epithelial cell penetration in mice, only the latter was capable of multiplying within host cells.

Studies with mutants of *S. typhimurium* defective in the LPS poly-saccharide antigen have shown that loss of the O antigens appears to increase sensitivity to antibody and complement. These mutants were also more easily phagocytosed than the parent, both *in vivo* as measured by clearance from the blood following intravenous injection, and as measured *in vitro*, using mouse peritoneal macrophages. Strains carry lesions deeper in the LPS core region also became sensitive to killing by lysosomal fractions of polymorphonuclear leucocytes. Those with a complete core were as resistant as the smooth parent. These observations appear to suggest that the complete core oligosaccharide is responsible for resistance to at least some of the bactericidal mechanisms within phagocytic cells.

Further evidence indicating that the O-antigen is an important determinant of resistance to ingestion by macrophages and virulence of *Salmonella* has come from the examination of genetically manipulated strains differing only in their antigenic side chain structure. Valtonen and co-workers compared strains of *S. typhimurium* bearing the LPS O specificity of *S. enteritidis* (O-9,12) *S. montevideo* (O-6,7), or the parent *S. typhimurium* (O-4-12) (Figure 5.10). They clearly showed that the strain possessing the O-6,7 antigen was least virulent while the parent serotype, O-4,12 showed the highest virulence. The strain with the O-9,12 antigen was of intermediate virulence. These results correlated well with the clearance rates of the strains, implying that the virulence was a reflection of resistance to phagocytosis. This was later confirmed by measuring uptake of the strains by macrophages *in vitro*. The rate of ingestion was found to be inversely proportional to virulence, and it was considered that the differences in uptake reflected differences in affinity of bacteria for macrophages rather than simply the rate of ingestion at saturating numbers of bacteria.

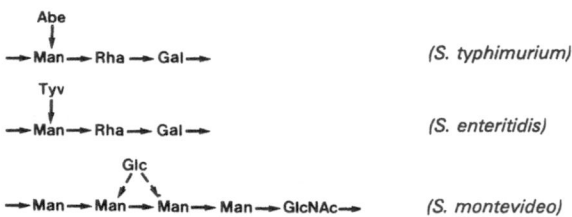

Figure 5.10: Structure of *Salmonella* O-repeating Oligosaccharides Shown to Differ in Their Contribution to Virulence

Because the uptake required complement, it was suggested that the process involved an interaction between the C3b receptor on the macrophages, and C3b on the bacterial surface. The quantity of C3b on the surface of the organism was thought to be determined by the ability of the polysaccharides present at the surface to activate complement by the alternative pathway. These findings imply that like capsular polysaccharides, some O antigens are more capable than others of protecting the cell by covering the bacterium with a surface layer which does not provoke activation of complement, thus avoiding both opsonophagocytosis and its direct bactericidal effects.

The shigellae present a different pathological picture. Organisms of this group penetrate and invade the epithelium of the colon and spread laterally causing the death of patches of intestinal tissue resulting in a bloody mucus diarrhoea. Invasion by these bacteria is thought to be limited to the intestinal wall through their susceptibility to ingestion and killing by phagocytic cells and the bactericidal activity of antibody and complement.

Of the four species only one, *Shigella dysenteriae*, is known to elaborate a toxin other than classical LPS endotoxin. This is a heat-labile neurotoxin (Shiga toxin) which also appears to possess enterotoxic properties. *S. flexneri* is the most widely studied member of the genus with respect to its pathogenic mechanism. It is now clearly established that the ability of this organism to penetrate the intestinal epithelium and multiply within the tissue is multideterminant, and that one such factor involved is the O specific polysaccharide. When the *S. flexneri* 2a O-side chain (group antigen) was replaced by introducing the genes determining the O8 antigen of *Escherichia coli* (a mannan), the recombinant strains were uniformly avirulent. They were unable to cause keratoconjunctivitis in guinea pigs, a test originally described by Sereny to assess the ability of an organism to penetrate epithelial cells of the cornea. A positive result in this test reflects the ability of an organism to invade the intestinal epithelium which is a necessary attribute for virulence in *Shigella*. Hence, it is not the presence of an O-side chain alone that is important, but the quality of that antigen. This finding was supported by concurrent experiments showing invasiveness to be retained when the O25 antigen of *E. coli* replaced the native *S. flexneri* group antigen. The structure of both the O25 antigen and the *S. flexneri* 2a group antigen includes the sugar rhamnose and it was suggested that the similarity of chemical composition between the two polysaccharides might explain the conservation of virulence by the hybrid strains.

Further studies with this organism showed that although the ability to evoke keratoconjunctivitis was lost in a strain lacking the O-side chain, it was still capable of invading tissue culture cell monolayers. This suggested that the O-side chain is not essential for cell penetration, but it may be important in the survival of the organisms during invasion *in vivo*. Recent work has clarified the determinants necessary for virulence in *S. flexneri*. These are reported to include a large (140×10^6 M_r) plasmid encoding the genes required for epithelial cell penetration (p. 179), and three chromosomal loci, one of which is the *his*-linked group 3,4 antigen determinant.

The type II specific antigen of *S. flexneri* 2a has been excluded as a virulence determinant. It consists of α-glucosyl secondary side chains attached to a rhamnose residue of the O-polysaccharide repeating unit. When crossed with *E. coli*, recombinants which had inherited the region of *E. coli* chromosome replacing the *Shigella* type II antigen determinants were shown to have lost this antigen but maintained their virulence.

Shigella sonnei, the least virulent member of the genus, comprises a single serotype. Smooth colonies possess an O specific polysaccharide termed the form I antigen which consists of a repeating disaccharide of two amino sugars: 2-amino-2-deoxy-L-altruronic acid and 2-acetamido-4-amino-2,4,6-trideoxy-D-galactose. The O specificity is spontaneously lost at a high frequency, and the resulting rough colonies which are always avirulent are termed form II. The form II antigen is the R1 LPS core oligosaccharide of enterobacteria. Recently, synthesis of the form I antigen has been shown to be directed by a 120×10^6 M_r plasmid, and this plasmid has been found to be essential for virulence. The direct involvement of the O-side chain moiety however is not established. Some transposon tagged derivatives of the plasmid were able to direct form I antigen expression, but the host bacteria were found to be avirulent indicating the presence of other, separate virulence genes on the plasmid. It may be found that determinants of epithelial cell penetration reside on the plasmid as in the large plasmids of *S. flexneri* strains. It has been suggested that the form I antigen may afford some degree of protection against immune factors at the site of invasion, however involvement of the form I antigen in *S. sonnei* dysentry awaits clarification by techniques capable of discriminating between different functions encoded on the plasmid.

To date 164 different O antigens are recognised within the single species

E. coli. Various groups of O antigens have been associated with certain types of *E. coli* infections in epidemiological studies, but little clear evidence pertaining to their involvement in systemic pathogenicity is available. One investigation showed the importance of the O-side chain O111 in the ability of the organism to resist phagocytosis. When present on the surface of the bacterium, the O111 antigen protected the strain from ingestion by phagocytic cells, and increased the mouse virulence of the strain up to 1000-fold over the same mutant strain grown under conditions in which it was unable to synthesise the O-polysaccharide.

As described previously the direct bactericidal activity of complement, both with and without specific antibody, is a major immediate defence against many Gram-negative organisms including *E. coli* and *Salmonella* spp. Strains devoid of O-specific peripheral polysaccharides are invariably rough and sensitive to serum complement, while those resistant, are usually smooth possessing an O antigen. Studies attempting to relate O-type to complement resistance have failed, as have efforts to show a correlation between the ratio of substituted to unsubstituted core stubs or the total amount of O antigen present, and complement resistance. Nevertheless, studies with smooth, insensitive strains and their *rfb* (O antigen-deficient) mutants clearly show the polysaccharide substituent of the LPS to confer some level of resistance. Problems encountered in interpreting the differing behaviour of unrelated clinical isolates probably reflect the multideterminant nature of this phenomenon, and it is likely that other surface-located elements are involved.

Evidence that exogenous LPS can protect organisms from the bactericidal activity of serum complement has been reported. Purified LPS from a strain of *E. coli* (ML 308 225) was found to bind to that strain, and relatively large quantities protected it from complement killing. Subsequently investigations found that only the purified LPS from the strain under examination could protect the cells; LPS from other sources showing little or no activity. The mechanism of resistance by the bound LPS was not understood, nor was the reason for the observed specificity of the reaction. The possibility that this phenomenon may have some importance during bacterial infection remains open.

The importance of the O8 antigen of *E. coli* ($\alpha 1 \rightarrow 3$ linked trisaccharides of $\alpha 1 \rightarrow 2$ linked D-mannose) was investigated in relation to serum complement resistance, by introducing the O8-polysaccharide determinants of a smooth form Hfr donor strain into its rough mutant recipient. The presence of the O8 antigen caused a delay in the bactericidal reaction, but did not fully protect the cell. It is clear that while the O-specific polysaccharides of *E. coli* and other Gram-negative bacteria can be important virulence factors, other determinants are necessary for the expression of full virulence.

Intestinal infection such as infantile gastro-enteritis, dysentery-like disease and travellers' diarrhoea are also caused by *E. coli*, which is usually considered to be a harmless natural inhabitant of the large bowel. Certain serogroups are associated with these infections, but the actual importance of the O antigen is not known. Cholera-like syndromes such as travellers' diarrhoea are clearly dependent upon the liberation of enterotoxins and the production of colonisation factors (p. 183), both of which are invariably plasmid mediated, by the so-called enterotoxigenic *E. coli*. The type of O-polysaccharide is not thought to be important in this type of disease, and this is reflected by the predominance of normal faecal serotypes among the toxigenic strains.

Enteroinvasive strains of *E. coli* appear to be very closely related to *Shigella* except for distinguishing biochemical traits. They have the same invasiveness as *Shigella* strains, and are positive in the Sereny test suggesting that they possess the same type of virulence determinants as the recognised pathogens. It is likely therefore that observations on the relation of *Shigella* LPS to its invasive capacity apply equally to these organisms. Only O-polysaccharides of certain undefined composition and structure can fulfill the necessary requirements of that antigen in the pathogenesis of the disease.

There are common examples of systemic bacterial disease where an O-polysaccharide is not required. *E. coli* isolated from the CSF of neonates suffering from meningitis frequently possess the K1 polysaccharide as noted previously (p. 150). They also usually possess one of a limited number of O serotypes namely O1, O2, O7, O18 and O78, or possess no O-side chain and are rough. This observation is particularly intriguing since all known examples of *Neisseria meningitidis*, the major cause of epidemic bacterial meningitis in older children and adults, also possesses no O-specific LPS, but instead R-type LPS with the same core and lipid A components as the Enterobacteriaceae. Whether the R-specificity plays some role in the pathogenicity of meningococcaemia such as tissue trophism, or its presence in both agents is merely coincidental, remains unclear. The LPS of *Neisseria gonorrhoeae* does appear to contain serologically distinct antigens analogous to O-polysaccharides. Although some investigations have been unable to detect O-side chains on gonococcal LPS, suggesting that like that of *N. meningitidis* it only occurs in the rough form, others have shown the O-side chains to be present on so-called type 1 colony variants but absent on type 4 strains. A possible role for the LPS in resistance of *N. gonorrhoeae* to the bactericidal activity of serum was suggested by the finding that loss of the LPS serotype antigens in a pyocin resistant strain was accompanied by increased serum sensitivity of the strain. As the pyocin utilises the LPS in binding to the cell, much like a bacteriophage tail, it was assumed that pyocin resistance was due to the LPS alteration alone, and therefore the complete LPS confers a degree of complement resistance upon at least some strains of *N. gonorrhoeae*. While not implicated in simple gonococcal infection, resistance to serum complement does appear to be an important feature of gonococci causing disseminated infection.

Amphiphilic Envelope Components Other than Lipopolysaccharide

Although less well studied than LPS, other bacterial amphiphiles are currently thought to be virulence factors in a variety of pathogens. Unfortunately there exists little definitive evidence concerning the nature of those structural features contributing to biological activity, however one common requirement for expression of almost all the biological properties is the presence of an intact hydrophobic lipid region. Envelope amphiphiles implicated in pathogenicity include the lipoteichoic acids found in nearly all Gram-positive bacteria, lipoproteins from Gram-negative organisms and the enterobacterial common antigen of bacteria belonging to the family Enterobacteriacae.

Lipoteichoic Acids (LTA). First isolated from *Lactobacillus fermentum*, lipoteichoic acids have now been found in many species of Gram-positive bacteria including pathogens and non-pathogens. Unlike other cell wall teichoic acids they show no covalent attachment to the peptidoglycan and have a lipid moiety at one end of the linear glycerophosphate polymer which is buried in the cytoplasmic membrane of the bacterial cell. The biological effects of LTA closely parallel those of endotoxic lipopolysaccharides except for lethal toxicity and pyrogenicity which are only associated with the latter. The structure of the lipid portion of the LTA molecule, analogous in biological activity to the lipid A of LPS, is not known in detail, but it has been shown to be a glycolipid with simpler lipid substituents than lipid A, and no ester-linked hydroxy fatty acids.

The involvement of LTA in the production of disease may be similar to that of lipopolysaccharide. LTA is released from the cell envelope during growth and has a strong affinity for membranes such as erythrocyte membranes. Group A streptococcal LTA is also known to activate complement in the absence of specific antibody. LTA from this organism has also recently been suggested to have a role in the binding of the cell to mucosal surfaces. The adherence previously attributed to the M-protein is actually due to LTA; the determinant of resistance to phagocytic ingestion being a separate structure, on the fimbriae-like layer, from that which binds to epithelial cells.

Enterobacterial Common Antigen (ECA). This envelope antigen common among many different genera of enterobacteria has received relatively little attention considering the potential diagnostic and prophylactic value of a common bacterial antigen. ECA can be detected in bacterial extracts by haemagglutination of erythrocytes exposed to the extract, in the presence of specific anti-ECA antibody.

Antiserum to ECA is not generally produced by the immunisation of animals with whole formalinised cells. This is because ECA is in a haptenic form, and unable to elicit an antibody response. Only in strains lacking an O-side chain where it is found linked to the LPS core oligosaccharide is it immunogenic. The ECA molecule consists of a lipid portion of poorly defined compositon linked to a linear polymer of 1→4 linked N-acetyl-D-glucosamine and N-acetyl-D-mannosamine-uronic acid.

In contrast to LPS and LTA, the biological properties of ECA are limited to its immunogenicity and an ability to bind to eukaryotic cell membranes. Membrane disruption with release of endogenous pyrogenic cellular constituents is not a feature of ECA binding. Evidence suggesting a link between possession of ECA by an organism and its pathogenic potential has come from examination of an ECA-negative (*rfe*) derivative of *Salmonella typhimurium*. When compared with its isogenic ECA-positive parent by intraperitoneal injection of mice, the mutant was found to have reduced virulence. The number of bacteria in the challenge, necessary to cause death in 50 per cent of the test animals, was shown to be 10-fold greater in the ECA-negative strain than the parent. Both of these strains were smooth (O-4,12), and it was suggested that ECA is required for full expression of the pathogenic potential in this organism. The means by which the ECA enhances virulence is not yet understood.

Lipoprotein (LP). Found in a wide variety of Gram-negative bacteria, it is the most abundant protein in the outer membrane of *E. coli*. The lipid portion of the molecule at the N-terminus of the polypeptide is associated with the inner leaflet of the outer membrane, the C-terminus often being covalently linked to the peptidoglycan. Like ECA the biological properties of isolated lipoproteins are limited, but in addition to the ability to bind eukaryotic membranes the lipoprotein possesses very potent mitogenic activity (p. 179).

Protein Components

Surface Protein Layers

Regularly arranged protein layers have been observed on a number of bacterial species and those present on pathogens appear to contribute to the ability of the organism to cause disease.

The fish pathogen *Aeromonas salmonicida* is the agent of a septicaemic disease of fish, primarily salmon and trout, termed furunculosis. Strains isolated from the infection are autoaggregating and have been shown to possess an additional protein layer known as the A layer or A protein. This

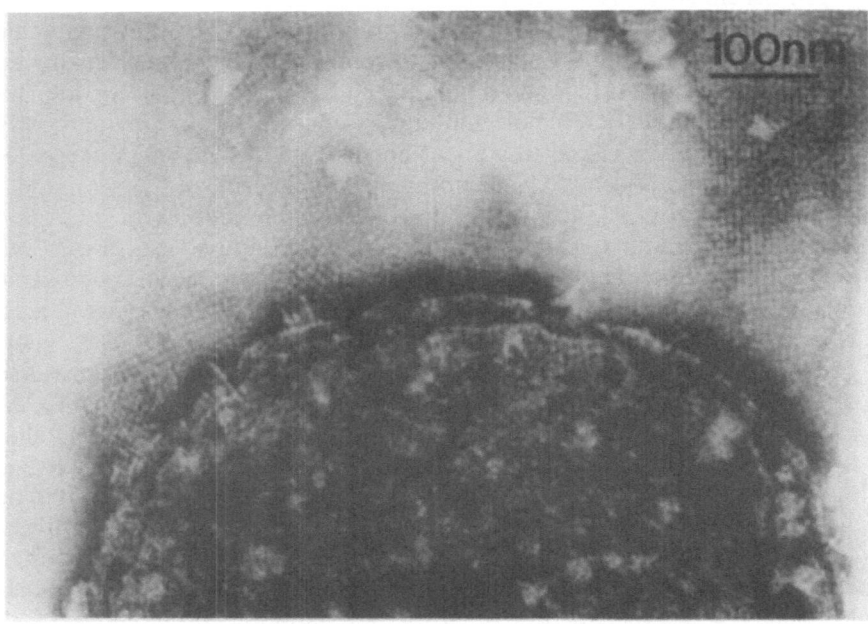

Figure 5.11: A Layer of *Aeromonas salmonicida* Separated from the Bacterial Cell. Note regularly ordered structure. Reproduced by permission: Kay, W.W., Buckley, J.T., Ishiguru, E.E., Phipps, B.M., Monette, J.P.L. and Trust, T.J. (1981) *J. Bacteriol.*, *147*, 1077.

layer has been found to form a tetragonal pattern, and is located at the surface of the cell (Figure 5.11). It is composed of single acidic poly-peptides of approximately 49,000 M_r which are rather hydrophobic. When the A protein is lost by mutation, it results in a dramatic attenuation of virulence, supporting a role for the protein in the pathogenesis of the infection. The mechanism by which the A protein enhances virulence is not clear but it appears to form a barrier, enclosing the whole cell, which is impermeable to large molecules while allowing the passage of small nutrient molecules and excreted wastes. It has been suggested that the layer may shield the organism from the host immune response, and in keeping with this idea is the finding that the protein does not elicit a detectable antibody response. Such a layer would be expected to reduce the suscepti-bility of the organism to phagocytosis and humoral host defences.

A similar surface protein layer has been found to play a role in virulence in the animal and human pathogen *Campylobacter fetus* subsp. *intestinalis*. This is also a hydrophobic protein, but unlike the A protein of *Aeromonas* its subunits are arranged in a hexagonal pattern and are larger (90,000 M). It was found to be the component responsible for the ability of this organism to resist phagocytosis.

Staphylococcal Protein A. Most strains of *Staphylococcus aureus*, but not the less virulent coagulase-negative straphylococci, contain a protein fraction known as protein A. This biologically active surface located constituent is covalently linked to the peptidoglycan, but the point of attachment is not clear. When released from the cell by heat or proteolytic enzymes it shows a molecular weight of about 13,000, however the molecule appears much larger (42,000 M_r) when removed using enzymes such as lysostaphin or lysozyme which act on the peptidoglycan structure.

The alternative name, agglutinogen A, reflects the properties of this protein. When originally described in 1958 it was found to give a serological reaction with all human sera. As much as 45 per cent of serum immunoglobulin was found to react with protein A, indicating that a normal, specific antigen/antibody reaction was not involved. In fact the protein A reacts non-specifically with the Fc portion of IgG from most mammalian sera, and not with the Fab (antigen binding) region. The molecule has a high binding affinity for human IgG-1, IgG-2 and IgG-4, it does not react with IgG-3 and has a low binding affinity for rabbit IgG. While it was previously thought to react solely with IgG, more recent evidence indicates that human colostral IgA and serum IgM are also bound non-specifically.

A role for protein A in the pathogenesis of staphylococcal infecton is not established, but evidence suggests that it competes with phagocyte membrane receptors for the Fc portion of opsonising antibody. A variable proportion of the protein A is released into the growth medium where it exerts an 'antiphagocytic' effect *in vitro*. Complement activation by IgG is not prevented by binding of protein A, and protein A itself activates complement by both pathways. The resulting generation of leucocyte chemotaxins (C3a and C5a) may partly explain the influx of leucocytes resulting in an abundance of pus which is characteristic of staphylococcal infections. A concomitant enhancement of opsonisation by C3 however, is not found. This is due to the blocking of heat-labile (complement derived) as well as the heat-stable (antibody Fc) opsonins by the protein A.

Outer Membrane Proteins

Although it is known that outer membrane proteins can influence the virulence of Gram-negative bacteria, few examples are documented. The outer membranes of *E. coli* and *Neisseria gonorrhoeae* have been examined in detail, and in both organisms evidence suggests involvement of the outer membrane proteins in the survival of the bacteria *in vivo*. A role for *E. coli* LPS-associated outer membrane proteins in the tissue damage previously attributed solely to classical lipid A endotoxin has also been found.

Comparison of the envelope proteins of a serum-resistant mutant of *E. coli* and its parent, showed an increase in 46,000 M_r protein in the mutant. It was therefore suggested that the greater quantities of this protein may be responsible for the observed resistance. The presence of complete LPS O-side chains appeared essential for expression of this resistance. No other phenotype could be assigned to the protein, but the ability of whole cells to evoke a strong antibody response to the protein, and the ease of extraction from cells indicated a surface location, probably in the outer membrane.

A major outer membrane protein of *E. coli* which is specified by plasmid located genes, has been shown to be responsible for resistance to the bactericidal activity of low concentrations of serum. Several conjugative plasmids have been found to confer this property, e.g. R100 and R6-5, and gene-cloning techniques have been employed to identify the gene responsible. This was found to be the *traT* gene whose product is an outer membrane protein involved in surface exclusion: the prevention of formation of stable mating pairs with strains carrying a conjugative plasmid of the same type. The protein of 25,000 M_r is known to be exposed at the surface of the outer membrane, and present in amounts corresponding to about 21,000 copies per cell. The mechanism by which complement resistance is achieved by *traT* is not known, nor is the contribution of that effect to virulence, but several reports have indicated that the effect is clearly enhanced when such a factor operates in the presence of other known determinants of virulence,˙ e.g. in the presence of a full complement of polysaccharide antigens.

Other plasmids have been associated with virulence in *E. coli*. The col V plasmid specifying production of, and immunity to colicin V is such a plasmid. For many years production of colicin itself was thought to confer virulence, but it has since been shown that the colicin and virulence determinants are distinct. Three virulence determinants are known: (a) the *traT* protein is produced and may be important in resisting serum complement (above). (b) A separate locus termed *iss* (for increased survival in serum) has been described with similar properties to *traT*, but the product has not yet been shown to be surface located. Although a new major outer membrane protein of 33,000 M_r from Col V−K94 has been demonstrated recently, there is no direct evidence as yet to connect this with *iss*. Investigation of strains carrying these determinants has found that they do not alter consumption of the terminal complement components. This suggests that they somehow interfere with the action of the membrane attack complex, rendering it unable to cause membrane damage rather than preventing its formation. (c) The third determinant specifies a system for enhanced iron uptake. The contributions of *traT* and *iss* carriage have been found not to be additive in resisting complement mediated bactericidal activity; the combined effects of enhanced iron uptake and complement resistance appear to explain the increased virulence of strains carrying this plasmid.

Virulence determinants necessary for epithelial cell penetration are located on a 140×10^6 M_r plasmid in *Shigella flexneri*. Although the nature of these determinants is essentially unknown, the plasmid (pWR110) has been shown to direct the incorporation of radiolabelled amino acids into some outer membrane proteins. It has been suggested that these polypeptides might modify the bacterial surface by functioning as receptors capable of binding to epithelial cells and somehow mediating endocytosis of the invasive bacteria.

Apart from the effects upon survival and growth within animal tissues, outer membrane proteins may also possess potent biological activity. It has been shown that the porins of *E. coli* (products of the *ompC* and *ompF* genes) are powerful mitogens and polyclonal activators of B lymphocytes, i.e. they non-specifically stimulate the blast transformation and proliferation of B lymphocytes in an animal. OmpA also has these properties, but to a lesser degree, and the lipoprotein is perhaps the most potent of these proteins. Endotoxin protein which is closely associated with the LPS after extraction under suitable conditions has been shown to be toxic for mice and to produce other diverse biological effects. Although it was suspected for some time that the observed activity was due to endotoxin contamination of these preparations, this has since been refuted. The endotoxin protein has now been identified as the known lymphocyte mitogens described above, i.e. the major outer membrane proteins, which possess biological activities distinct from the LPS.

Outer membrane proteins of *N. gonorrhoeae* have been found to undergo selection in guinea-pig subcutaneous chambers (p. 151). Strains possessing certain outer membrane proteins were found to survive the highly competitive conditions while others were lost. The likely implication is that the protein composition of the outer membrane is an important factor for survival of the gonococcus *in vivo*, and this is probably a reflection of resistance to host defences. The persistence of pilated, opaque strains within the chambers indicates a role for the structural proteins associated with these forms in the pathogenesis of gonorrhoea, but details of that role remain unknown.

Surface Appendages and Microbial Virulence

The adherence of bacteria to mammalian tissue surfaces may be an important event in the establishment of many bacterial infections. Mucosal and endothelial surfaces are constantly bathed in fluids, e.g. mucus, blood, urine. These secretions are often kept in motion by a variety of anatomical mechanisms serving to cleanse the surfaces, e.g. sneezing, coughing, ciliary action or peristalsis. The ability to adhere to epithelial surfaces is thought to permit colonising bacteria to resist the secretory flow, which would ultimately sweep the organism from the body. Adherence of the pathogen must be followed by tissue invasion, either directly or by the bacterium releasing tissue-damaging toxins. However it is not yet fully established, except in a few specific cases, that bacterial adherence is a prerequisite for infection, and in no case does adhesion alone induce disease. The ability to bind to mammalian cells is a common property of bacterial fimbriae, and many studies indicate that these structures play a significant role in microbial virulence by facilitating the adhesion of the pathogen to the target tissue. Flagella do not mediate adherence and are generally not thought to be important factors in microbial pathogenicity.

Many Gram-negative clinical isolates possess long straight proteinaceous filaments around the cell periphery, collectively known as fimbriae. A majority of strains of *E. coli* and many strains of *Shigella, Klebsiella* and *Salmonella* species possess type I fimbriae (p. 140), defined in terms of their morphology and haemagglutinating specificity for red blood cells of different species. Strains bearing type I fimbriae readily agglutinate the erythrocytes of most animal species, with the exception of the ox. The haemagglutinating property is best expressed in bacterial cultures grown in nutrient-rich broths, in static culture after 48h. Examination of cultures grown in this way often reveals a 'skin' or pellicle at the air/liquid inter-face, containing large numbers of fimbriate bacteria. The haemagglutin-ating properties of organisms bearing type I fimbriae are inhibited by the addition of small concentrations of D-mannose or mannose-related compounds, e.g. methyl-α-D-mannoside and yeast mannan. Other sugars or sugar polymers have little inhibitory action. Modification of the hydroxyl groups at C-2, C-3, C-4 and C-6 positions of the D-mannopyranosyl ring diminishes the inhibitory response, suggesting that these groups are important in the interaction. The D-mannose serves as a soluble analogue of fixed D-mannose-like residues located on the surfaces of cells, to which bacteria containing type I fimbriae adhere. The sugar binds to a mannose-specific receptor(s) on the type I fimbriae, preventing the organelle attaching to the mammalian cell surface.

Although they share the property of mannose sensitivity, the type I fimbriae present in different enterobacteria are not identical in composition. Type I fimbriae among *E. coli, Shigella flexneri* and *Klebsiella aerogenes* possess limited antigenic homology. Similarly some antigenic homology can be detected in *Salmonella, Citrobacter* and *Arizona* species type I fimbriae, but this latter group shows no cross reaction with the former. A third antigenic form of type I fimbriae is present in the genera *Edwardsiella, Enterobacter, Serratia, Hafnia* and *Providencia*. Since all these species bear a common mannose-sensitive adhesive site, it suggests that the sites are not sufficiently numerous or antigenically potent, to raise antibodies, and that the antigenically dominant region of the organelle is the fimbrial protein. This leads to the hypothesis that the mannose sensitive sites occupy a limited area of the organelle, presumably at the distal tip. In the environment the fimbriae radiate from the bacterial surface as a result of the mutual repulsion of their lateral surfaces, which must be sufficiently hydrophilic to prevent them binding together. The tip of fimbriae therefore face outwards ensuring that the adhesive sites present at the tip can make contact with mammalian cells. Type I fimbriae from *E. coli* and *N. gonhorroeae* can attach by their tips to almost any hydrophobic surface. The fimbria tip is thought to be the point of initial contact and to be able to penetrate the electrostatic barrier by means of its small surface area and the presence of uncharged hydrophobic residues. This would tend to reduce the electrostatic repulsion of the fimbria, as compared to the bacterial cell surface. A hydrophobic region on the fimbrial tip could also become embedded in the lipid interior of host cell membranes. It is thought that hydrophobic interactions of this kind are concerned with the initial stages of fimbrial binding, producing a preferred alignment of the organelle, which facilitates the later and stronger receptor-specified interactions.

Circumstantial evidence suggests that type I fimbriae may assist bacteria to colonise the intestine or urino-genital tract by facilitating the attachment of the organism to epithelial surfaces, thereby resisting the scouring action of body secretions. It must be stressed however, that the possession of fimbriae is not a prerequisite for survival within the host; many pathogenic and commensal bacteria are neither fimbriate nor adhesive, yet maintain substantial populations within the body. Many strains of enteric bacteria capable of causing intestinal infection, are invariably non-fimbriate. In these species the O-side chain of the LPS may mediate a weaker-type of adhesive interaction, capable of inducing uptake by host cells. However in many other species there is a strong correlation between the possession of fimbriae and pathogenic potential. Most *Salmonella* species capable of infecting man and animals bear mannose-sensitive fimbriae. When attempting to establish experimental infections in mice with either fimbriate and otherwise isogenic non-fimbriate strains of *Salmonella typhimurium*, greater success is achieved with the former. About three-quarters of *E. coli* strains isolated from the gut possess type I fimbriae. Similarly the majority of *E. coli* isolates from patients with urinary tract infections are fimbriate and bind strongly to urinary tract epithelial cells (however a significant proportion are neither fimbriate nor adhesive).

Significant changes in the adhesive properties of an organism may occur during the pathogenic process. *Proteus mirabilis* can cause ascending urinary tract infections in rodents. Adjustment of the growth conditions *in vitro* can produce *P. mirabilis* cells that are either heavily or lightly fimbriate. The former bind strongly to mammalian cells, but the latter adhere poorly. When injected into the rat bladder heavily fimbriate organisms are more likely to produce renal infection than lightly fimbriate organisms. However if the organisms are introduced intravenously, the fimbriate organisms are readily phagocytosed and the infection is cleared within 5 hours, while non-fimbriate organisms, being much less susceptible to phagocytosis, reach the kidney and establish an infection. During natural infection by *P. mirabilis* it is thought that the degree of fimbriation becomes modified as an adaptation to the changes in the nature of the environment as the organism penetrates the kidney. In the renal pelvis, the bacteria need the adhesive properties of the fimbriae to prevent them being swept away by the urine flow. On gaining access to deeper tissues, the possession of fimbriae becomes a liability, since they predispose the organism to phagocytosis, and in consequence the number of fimbriae declines. Similarly although meningococci isolated from the pharyngeal cavity of chronic human carriers of the organism adhere well to isolated epithelial cells, bacteria isolated from the blood stream are generally non-adhesive. It appears that the organism requires fimbriae to maintain its population in the pharynx, but on entering the bloodstream, the production of fimbriae is repressed to protect the organism from phagocytes.

Many *E. coli* isolates, plus strains of *Erwinia, Salmonella* and *Klebsiella*, possess fimbriae which show a different pattern of haemagglutinating specificity than strains bearing type I fimbriae. In addition their haemagglutinating properties are not inhibited by D-mannose. The adhesive strength of mannose-resistant fimbriae shows a marked temperature dependence; being strongest at 4°C and declining with temperature, until all activity is lost above 40°C. Certain enterotoxigenic strains of *E. coli* (ETEC) are associated with diarrhoeal diseases in man and young domestic farm animals. Although many differences exist between ETEC strains, they generally share three common properties. (a) They exhibit great avidity for the mucosal epithelium of the gut. This allows the organism to resist the flushing action of peristaltic movements of the intestine associated with diarrhoea, allowing the growth of a large bacterial population. Although the presence of fimbriae or fimbriae-like organelles has not been demonstrated in all ETEC strains, the K88 and K99 adhesins responsible for diarrhoea in pigs or calves respectively, possess surface structures closely resembling type I fimbriae. However adhesion by K88 and K99 strains is not inhibited by mannose. ETEC strains from humans usually bear equivalent structures termed colonisation factor I or II (CFAI/II). (b) ETEC strains produce enterotoxins, that act upon the membranes of the mucosal epithelial cells of the small intestine, causing the release of fluid into the lumen of the gut to produce the clinical symptoms of the disease. ETEC strains produce either singly or in combination, a heat stable (ST) or heat labile (LT) toxin. LT produced by *E. coli* resembles cholera enterotoxin and activates adenylate cyclase in the epithelial cell membranes, eliciting a transmembrane efflux of water and certain ions into the gut lumen. (c) The production of K88, K99 and CFA is controlled by transmissible plasmids.

There is much experimental evidence that supports a role for K88 and K99 fimbriae in promoting intestinal infection since suitable experimental models can be developed in piglets, calves and lambs. The evidence for CFA I and II is mainly circumstantial or derived from epidemiological studies, since no suitable animal model exists. The K88 adhesin is produced within the small intestine and its production permits K88 positive strains to adhere to and colonise that site but not the large intestine. Loss of the plasmid controlling K88 synthesis results in strains unable to produce the K88 adhesin or colonise the intestinal mucosa or cause diarrhoea. The re-introduction of the K88 plasmid into the K88-negative strains simultaneously restores the K88 antigens, the adhesive properties and the clinical symptoms. Piglets phenotypically lacking the receptor for the K88 fimbriae, or suckled by dams receiving K88 vaccines are resistant to infection by K88-positive ETEC strains. Similarly experiments in calves and lambs show that the K99 fimbriae are essential for colonisation of the small intestine and production of diarrhoea in the specific host. The vaccination of pregnant animals with purified K99 antigen, prevents diarrhoea in subsequent offspring when challenged with K99-positive ETEC strains. Examination of enterotoxigenic strains of *E. coli* from human adults suffering from diarrhoea, shows that the great majority (in some cases up to 98 per cent) possess one or other of the colonisation factors (CFAI or II), suggesting that these fimbriae are the human equivalents of K88 and K99. Loss of the plasmid controlling CFAI or II causes the strain to lose its ability to colonise the small intestine of infant rabbits. When human volunteers are infected orally with CFA$^+$ or CFA$^-$ bacteria, clinical symptoms are only observed in those receiving strains bearing CFA. The mannose-resistant fimbriae of ETEC strains exhibit binding to an extremely restricted range of host tissues. The K88 and K99 strains do not cause diarrhoea in humans and CFA$^-$ strains have little effect in calves or pigs. The intestinal receptors for these adhesive elements are not yet fully defined but may be β-D-galactosyl residues in heterosaccharide chains of membrane glycoproteins.

The initial symptoms of cholera resemble those of ETEC diarrhoea: there is accumulation of fluid in the lumen of the intestine, as a result of the activation of adenylate cyclase in the intestinal mucosa by cholera enterotoxin. Large numbers of vibrios can be detected in the spaces between the intestinal villi. Organisms appear to be able to penetrate the mucus that lines the intestine and reach the crypts at the base of the villi. The prevalence of cholera in communities possessing contaminated drinking water, with its associated high mortality, particularly in infants, has prompted many studies of the organism and the host, in an attempt to explain the complex interaction of the cholera vibrio with the surface of the intestine.

Lining the gut is a layer of mucus, serving to isolate the intestinal epithelial cells from the lumen contents. The oligosaccharide components of this gel differ in composition, not only at different points along the gut, but also between the tips of the villi and the basal crypts. The strains of *Vibrio cholera* best able to colonise the human gut and produce the clinical symptoms of cholera, can be distinguished from the many other vibrios present in water by the possession of a cell wall antigen, designated O1. The other vibrios, commonly known as water or non-cholera vibrios, produce cell wall antigens other than O1 and although not causing classical cholera, may produce cholera-like conditions. Virulent cholera vibrios exhibit strong haemagglutinating properties, presumably a reflection of the affinity of the organism for intestinal epithelial cells. The addition of D-mannose may inhibit the haemagglutinins of *V. cholerae* totally, partially or not at all, depending upon strain. The major haemagglutinin produced by classical strains of *V. cholerae* is not mannose-sensitive but is inhibited by the addition of L-fucose. This and other evidence, suggests that strains of *V. cholerae* produce several distinct haemagglutinins, often on the same organism.

Motility plays a significant role in the virulence of cholera vibrios. Non-motile mutants have been shown to be less able to bind to epithelial cells or produce cholera-like syndromes in experimental animal models, than their motile parents. The vibrios appear able to detect and move towards the mucosa in response to a chemoattractant gradient. The vibrios penetrate and then pass through the mucous gel. In order to reach the epithelial cells the vibrios must be motile to be able to drive the organism through the viscous mucus. In addition they must be able to sense the gradient of stimuli diffusing from the crypts, so that they migrate in the appropriate direction. The flagella of vibrios are sheathed. Vaccination with naked flagella provides little protection against an oral challenge with virulent vibrios in infant mice, whereas previous injection with sheathed flagellar preparations are effective.

The apparent ease with which cholera vibrios migrate through the intestinal mucus suggests that there is little permanent adhesion of the bacterium to this layer. Microscopical examination of cholera-infected rabbit intestine reveals large numbers of vibrios attached to the brush border surfaces of intestinal epithelial cells, but no evidence of fimbriae can be seen. There appear to be at least two highly specific receptors for the cholera vibrio on the mucosal surface, one of which is fucose-sensitive and located on the brush border epithelium. The location of the second fucose-resistant binding site is unclear. *V. cholerae*, like many other vibrios, produce a protein-rich slime capsule. There is no evidence that this layer has any adhesive properties, indeed the reverse may be true, since slime production and haemagglutinating activities are inversely related.

The cholera vibrio is non-invasive, and appears to be confined to the lumen of the gut. Since the cholera toxin acts directly upon the epithelial cell membrane it can be argued that a close association between the vibrio and the brush border of the epithelial cell would provide optimum conditions for the delivery of the toxin to its active site. It is an attractive, though not proven, hypothesis that the haemagglutinins located on the vibrio surface mediate this interaction.

The mucosal surface of man and mammals is the natural habitat of bacteria of the genus *Neisseria*. This group of organisms contains many common commensal species and two important pathogens: *N. gonorrhoeae* and *N. meningitidis*. Although *N. meningitidis* may invade the body to produce life threatening septicaemia and meningitis, it is a relatively rare event. This is surprising, since the carriage rate for this organism in the nasopharynx may reach 10 per cent of the population. This leads to the belief that both meningococci and gonococci are primarily pathogens of epithelial surfaces. Since both infections are confined to man, it appears that the organisms possess a high degree of adaptation to the human mucosal surface.

Fimbriae are universally present on gonococci and meningococci when freshly isolated. After sub-culture non-fimbriate variants soon outgrow the fimbriate gonococci indicating that growth within the host exerts selective pressure in favour of fimbriate organisms. In human volunteers fimbriate strains readily produce the clinical symptoms of gonorrhoea whereas non-fimbriate strains are avirulent. Fimbriate gonococci not only bind in greater numbers to isolated human cells, but also show an increased attachment rate over non-fimbriate bacteria. This rapid adhesion may be critical in the transmission of the infection, since the flow of body secretions may quickly remove the bacterium from mucosal surfaces. Antibodies to fimbrial proteins can be detected in the serum of patients suffering from gonorrhoea, suggesting that fimbriae are expressed *in vivo*. Immunisation with purified fimbriae increases the resistance of human volunteers to experimental gonococcal infection. Tests *in vitro* have shown that the possession of fimbriae increases the avidity of gonococci for human fallopian tube epithelial cells or human foreskin tissue, but not for human bronchial mucosa or leucocyte cells derived from the urinogenital tract of pig, rabbit or cow. This suggests that the gonococcal fimbriae are not only capable of discriminating between tissue types but also between species. An important indicator to the nature of the tissue receptor for the gonococcal fimbriae is the observation that low concentrations of gangliosides (sialic acid-containing glycosphingolipids) inhibit adhesion of the bacterium to human cells. Since gangliosides are known to be present on the surface of epithelial cells, it is thought that the receptor of the gonococcal fimbriae is likely to be a ganglioside-type molecule. Analysis of the inhibition of gonococcal fimbrial adhesion by a range of gangliosides or ganglioside

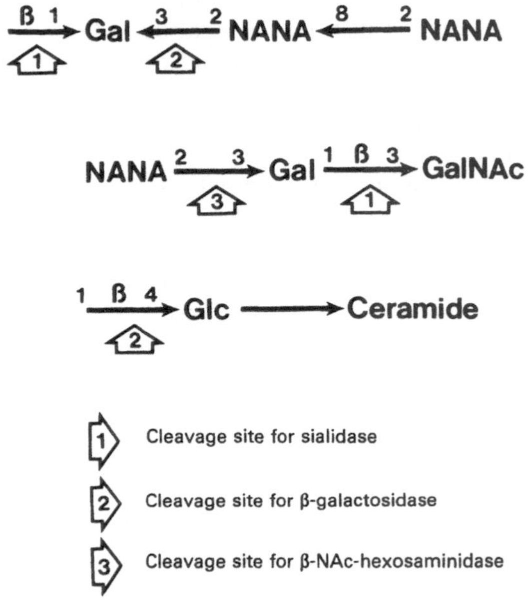

$$\xrightarrow{\text{ß 1}} \text{Gal} \xleftarrow{\text{3} \quad \text{2}} \text{NANA} \xleftarrow{\text{8} \quad \text{2}} \text{NANA}$$

$$\text{NANA} \xrightarrow{\text{2} \quad \text{3}} \text{Gal} \xrightarrow{\text{1 ß 3}} \text{GalNAc}$$

$$\xrightarrow{\text{1 ß 4}} \text{Glc} \longrightarrow \text{Ceramide}$$

⟨1⟩ Cleavage site for sialidase

⟨2⟩ Cleavage site for β-galactosidase

⟨3⟩ Cleavage site for β-NAc-hexosaminidase

Figure 5.12: Ganglioside Structures and Enzymes Cleavage sites. After Pearce and Buchanan (1980)

fragments generated by limited enzymic hydrolysis (Figure 4.12) has been used to identify the receptor. Sialic acid may constitute part of the receptor at pH 4.5 since at this pH only the sialic acid moiety can prevent attachment of *N. gonorrhoeae* fimbriae to human buccal cells. This binding may be the result of simple charge attraction, since at pH 7.5 sialic acid residues appear incapable of inhibiting the binding reaction. Treatment with β-galactosidase suggests that the terminal galactose residues are important in fimbrial adhesion at both pH 4.5 and 7.4. Treatment with β-hexosaminidase produce no effect. These data suggest that at pH 4.5 the receptor may contain NANA2→3 Gal/NAcβ1→4 Gal, but at pH 7.4 a Galβ1→3 GalNAcβ1→4 Gal configuration may be required. Treatment of buccal cells with a variety of hydrolytic enzymes, including xylosidase, mannosidase, fucosidase or glucosidase, in addition to sialidase, β-galactosidase and βNAc-hexosaminidase did not produce additional inhibition of attachment, consistent with their respective substrates not being part of the fimbrial receptor site. In contrast with type 1 fimbriae, which exhibit little selectivity in the type of cells to which they adhere, the gonococcal fimbriae selectively bind to a restricted number of tissues in a specific host. Whether or not the specificity of gonococcal adherence reflects differences in ganglioside composition in the tissues concerned is not yet clear.

Corynebacteria are the sole Gram-positive group of organisms that possess fimbriae and these also appear to mediate adherence to host tissues. Electron micrographs show that fimbriate *Corynebacterium renale* cells adhere strongly to cultivated mammalian cells, while non-fimbriate bacteria or bacteria treated with antisera raised against isolated fimbriae are unable to bind. The possession of fimbriae increases the virulence of *C. renale in vivo*. If the bladders of mice are inoculated with either fimbriate or non-fimbriate strains of *C. renale*, the former produce a greater incidence of pyelonephritis, ureteritis and cystitis. Scanning electron micrographs of the infected mouse bladder show large numbers of fimbriate bacteria adhering to the epithelial cells.

Group A *Streptococcus pyogenes* are common human pathogens responsible for many throat infections, including tonsillitis. This Gram-positive organism is able to adhere to and ultimately colonise the epithelial cells of the pharyngeal region. Electron micrographs suggest that the surface of *S. pyogenes* is covered with a layer of extremely fine projections that superficially resemble fimbriae. This surface layer is antigenic, containing the type specific M antigen (p. 38). The M proteins can be extracted under acid conditions yielding a complex mixture of partially degraded polypeptides (M_r 20,000 to 40,000) or by milder treatments to yield higher-molecular-weight proteins. Other surface components, including an M-associated protein, hyaluronic acid capsule and lipoteichoic acid appear to be intimately associated with M-protein to produce the fimbrial-like organelle. The M protein complex is thought to be important in the binding of *S. pyogenes* cells to epithelial cells although other factors are also thought to be involved. The exact nature of the M protein complex binding site is not clear. Pepsin treatment selectively removes M proteins from the surface of *S. pyogenes* cells, however this does not appear to affect the binding of the bacterium to isolated epithelial cells. Small amounts of lipoteichoic acid are exposed on the surface of group A streptococci and there is some evidence to suggest that the ester-linked fatty acids of the lipoteichoic acid are involved in the attachment process.

Role of Surface Components in the Virulence of Mycobacteria

Mycobacteria, in addition to being responsible for human tuberculosis and leprosy, are important pathogens in other animals. Unlike most other bacteria, mycobacteria are able to survive ingestion by white cells and successfully parasitise the host. The extent to which each species is present within the phagocyte or extracellularly varies amongst mycobacterial diseases and the host species, for example in severe cases of human pulmonary tuberculosis caused by *Mycobacterium tuberculosis* the bacteria are extracellular. In contrast when the same organism infects mice the bacteria are intracellular. The leprosy bacterium, *M. leprae*, is almost invariably intracellular. After ingestion by the white cell the mycobacteria grow and divide. The indifference of mycobacteria to the normally highly effective antibacterial action of phagocytes has prompted intensive study of the role of surface components in the pathogenesis of the disease. In order to survive within phagocytes the bacteria must not only be able to resist the intracellular killing mechanisms of the white cells but also fail to provoke any cytotoxic responses that could lead to the premature demise of the host cell. Although it is by no means clear as to how the bacterium accomplishes this difficult feat, the nature of the interface between the bacterial surface and the host cytoplasm must play a central role.

The cell walls of mycobacteria characteristically contain significant quantities of glycolipids; some intimately associated with the peptidoglycan layer and some existing free. In *Mycobacterium bovis* BCG, for example, about one third of the lipid content of the cell wall is readily extracted with organic solvents, the remainder being covalently linked to the peptidoglycan. In essence the composition of the peptidoglycan of mycobacteria resembles that of *Bacillus* species (p. 6), excepting that the acylation of the amino group of muramic acid residues is by glycolyl rather than acetyl groups. The cross-linking of the peptidoglycan strands is also rather unusual in that although some of the bridges resemble those present in *Bacillus* and *E. coli* (i.e. from D-alanine of one pentapeptide chain to meso-diaminopimelic acid in a second chain; Figure 1.2), other direct links between two diaminopimelic acid residues on adjacent chains have been described. Present at the C-6 phosphate residue of muramic acid of mycobacterial peptidoglycan is a large branched polysaccharide, a D-arabino-D-galactan, which constitutes the major surface antigen of the species (Figure 5.13). The wall-bound glycolipids, termed mycolic acids, are esterified to the arabinogalactan. The mycolic acids are long chain (C_{50}–C_{90}) fatty acids branched at the α, and hydroxylated at the β-position. Examination of mycolic acids from many species indicates that the acids commonly contain unsaturated regions, additional methyl substituents and cyclopropane rings.

In addition to the wall-bound mycolic acids three other important lipids are loosely associated with the murein layer. The first of these is the 'cord factor', so called since it was associated with the formation of 'cords', parallel rows forming characteristic serpentine strands demonstrable when smears of liquid grown cultures of mycobacteria are examined under the microscope. The 'cord factor' of *M. tuberculosis* can be readily extracted from viable bacteria with organic solvents and identified as trehalose-6,6'-dimycolate. The mycosides constitute the second group of readily extractable lipids, gycosidically linked to the *para* position of a phenol .

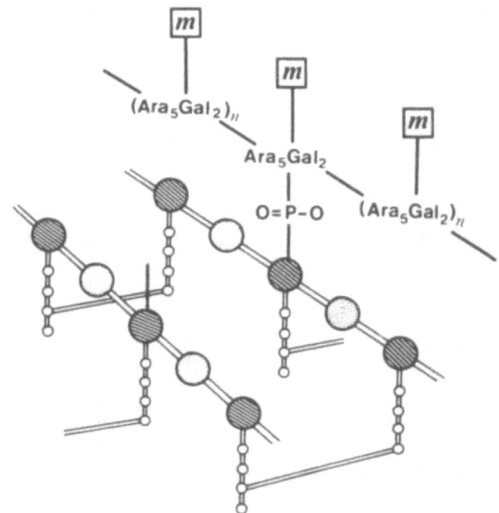

Figure 5.13: Representation of the Structure of the Mycobacterial Cell Wall. (Symbols as in Figure 1.2.) An arabinogalactan polymer $(Ara_5Gal_2)_x$ is linked to the C-6 phosphate of N-acetyl-muraminic acid and to mycolic acid (m). For simplicity only one of the peptidoglycan chains is shown linked to arabinogalactan.

ring, substituted by a long-chain branched alkyl residue. A third group of free lipids extracted from the walls of virulent tubercule bacteria, known as sulphatides, have been shown to consist of multiacylated trehalose sulphates.

Also present in the walls of *Mycobacteria* are significant amounts of glucan and poly-L-glutamic acid. The latter may have a molecular weight of up to 35,000 and constitute up to 10 per cent of the weight of the wall.

Many attempts have been made to identify the role of specific myco-bacterial wall components in determining the pathogenic potential of individual strains. Much of the data is equivocal and often controversial, but some limited correlations can be drawn. The 'cord factor' seems capable of inducing swelling and disruption of mouse liver mitochondria and may have an inhibitory effect on the production of hydrolytic enzymes by phagocytic white blood cells. In the normal course of events the ingestion of bacteria by phagocyles results in the formation of a vacuole or phagosome. Within the white cell powerful hydrolytic enzymes and bacteri-cidal proteins are localised in small vesicles termed lysosomes. These fuse with the phagosome releasing their contents into the interior of the vacuole. Isolated mycobacterial sulphatide has been shown to be a powerful inhibitor of lysosome-phagosome fusion. It has also been pointed out that virulent strains of mycobacteria possess strongly acid surface lipids, in particular sulphatides.

Table 5.1: Vaccines Used in Prevention of Some Bacterial Diseases in Humans

Disease	Vaccine composition	Applications
Cholera	Crude fraction from *Vibrio cholerae*	people living and working in endemic areas
Diphtheria	Purified toxoid	
Tetanus	Purified toxoid	infants, boosters for adults at risk
Whooping cough (Pertussis)	Killed *Bordetella pertussis*	
Tuberculosis	Live *Mycobacterium bovis* (BCG) an attenuated strain	all tuberculin-negative persons
Typhoid and *Paratyphoid fever*	Killed *Salmonella typhi* and *Salmonella paratyphi*	people living and working in endemic areas or areas suffering an outbreak
Pneumococcal pneumonia	Purified polysaccharides from *Streptococcus pneumoniae*	adults with chronic systemic diseases
Meningococcal meningitis	Purified polysaccharides from *Neisseria meningitidis*	people at high risk
Plague	Crude fraction from *Yersinia pestis*	workers at risk from wild rodents in endemic areas

Adjuvants enhance the immunological response of animals and if injected into experimental animals with an antigen dramatically increase the titre of antibodies produced. Preparations containing mycobacterial cell walls in an oil/water emulsion (known as Freund's adjuvant) are extensively used by immunologists. Similar mixtures which do not contain mycobacteria promote a poor response. The stimulatory effect of the mycobacteria appears due to a fragment of the peptidoglycan, N-acetylmuramyl-L-alanyl-D-isoglutamyl-meso-diaminopimelic acid. In addition to being a powerful adjuvant this fragment also enhances the non-specific immunity of the host to Gram-negative bacteria.

Infection with mycobacteria appears to stimulate the immunity of the host to tumours. Isolated cell walls also provide some antitumour activity but treatment with proteolytic enzymes or organic solvents destroys the tumour-suppressive properties of the walls. Subsequent addition of 'cord factor' restores this activity.

Exploitation of Cell Envelope Components on Protective Vaccines

It is hoped that a better understanding of the role of bacterial envelope components as determinants of pathogenicity will lead to the development of improved vaccines. Careful selection and combination of antigens should provide vaccines which afford protection against infection by neutralising one or more of the virulence determinants, and so interfere with the pathogenesis of the disease.

To date most vaccines have been developed successfully on an empirical basis. Table 5.1 lists the most important vaccines used in the prevention of bacterial diseases in man. Diphtheria and tetanus vaccines are toxoids, inactivated toxins which give immunity by production of antibody directed towards the lethal toxins released by the organisms upon infection. The other vaccines contain envelope components which give immunity by production of antibodies directed against surface antigens. Infection is thus prevented at an early stage before serious damage to the host. The antigen preparations range in complexity from purified capsular polysaccharides (e.g. pneumococcal vaccines) to crude extracts of whole cells (cholera vaccine), killed whole cells (e.g. whooping cough vaccine), and finally to live, attenuated whole cells (e.g. BCG tuberculosis vaccine).

The precise nature of the protective components of the crude vaccines is unknown and the overall protection probably results from a combination of antigens. Nevertheless, the success of purified antigens is encouraging for the future rational design of vaccines. Many envelope components identified as important determinants of pathogenicity are being investigated for protective activity, especially capsular polysaccharides, lipopolysaccharides, outer membrane proteins and fimbriae.

Endemic bacterial meningitis in infants caused by *Haemophilus influenzae* type b (HIb) is one disease which might be effectively controlled by vaccine prophylaxis. Despite antibiotic therapy the disease has a mortality rate of about 10 per cent with an incidence of 0.5 per cent in the USA. The isolated polysaccharide capsule of HIb (phosphoribosylribitolphosphate) has protective activity upon immunisation, especially when traces of protein and lipopolysaccharide are included. Similar capsular polysaccharides are found in *E. coli* K1 and group B meningococci so it might prove feasible to devise a common vaccine. There are other examples of antigenically similar capsular polysaccharides occurring in different bacteria (e.g. certain *E. coli, K. aerogenes,* and *S. pneumonia* strains) so broad spectrum vaccines might one day be developed.

Attention has been given to exploitation of the protective properties of fimbriae preparations. Fatal diarrhoea (scouring) of piglets and calves, especially under intensive rearing can be controlled by vaccines containing the fimbriated K88 and K99 serotypes of *E. coli* which are responsible for the infections. Antibody against the adhesive fimbriae prevents colonisation of the intestine of the neonatal animals by K88 and K99 enterotoxigenic strains. A similar approach might be used to control the fatal diarrhoea of children in some undeveloped countries caused by enterotoxigenic *E. coli* bearing the CFAI and II fimbriae.

Gonorrhoea might be controlled at the colonisation stage using vaccines containing fimbriae or, possibly, the principal outer membrane proteins which are also important in the adhesion and colonisation of *Neisseria gonorrhoeae.*

In addition to the development of new vaccines for the control of a wider range of bacterial infections there is plenty of scope for improving existing preparations. Adverse reactions must be eliminated and, for many vaccines the duration of protective cover needs to be extended. A key factor in any improvements will be a better understanding of the nature, properties and function of vaccine components. Genetic engineering has a vital role to play, both in studying potential protective antigens and in maximising their production.

Further Reading

Babior B.M. 'Oxygen-dependent Microbial Killing by Phagocytes', *New England Journal of Medicine* (1978), 659-68 and 721-44

Beachey, E.H. *Bacterial Adherence* (Chapman and Hall, London, 1980)

Bitton, G. and Marshall, K.C. (eds) *Adsorption of Microorganisms to Surfaces* (Wiley Interscience, New York, 1980)

Draper, P. 'Mycobacterial Inhibition of Intracellular Killing' F. O'Grady and H. Smith (eds), (Academic Press, 1981)

Jann, K. and Westphal, O. Microbial Polysaccharides', In M. Sela (ed.) *The Antigens*, vol. 3 (Academic Press, London, 1975) p 1–125

Lachmann, P.J. 'Complement', In M. Sela (ed.) *The Antigens*, vol. 5 (Academic Press, London, 1979) p. 283–335

Mims, C.A. *'The pathogenesis of infectious disease*' (Academic Press 1982)

Orskov, I., Orskov, F., Jann, B. and Jann, K. 'Serology, Chemistry and Genetics of O and K antigens of *Escherichia coli, Bacteriological Reviews* (1977), *41*, 667–710

Pangburn, M.K. 'Activation of Complement via the Alternative Pathway', *Federation Proceedings* (1983), *42*, 139–42

Pearce, W.A. and Buchanan, T.M. 'Structure and cell-membrane-binding properties of bacterial fimbriae' In E. H. Beachey (ed.), *Bacterial Adherence* (Academic Press, London, 1980) pp. 289–344

Smith, H. 'Microbial Surfaces in Relation to Pathogenicity', *Bacteriological Reviews* (1977), *41*, 475–500

Smith, H., Skehel, J.J. and Turner, M.J. *The Molecular Basis of Microbial Pathogenicity*, Report of the Dahlem workshop on the molecular basis of the infective process (Chemie Verlag, Weinheim, Federal Republic of Germany. 1980)

Spitznagel, J.K. 'Oxygen-independent Antimicrobial Systems in Polymorphonuclear Leukocytes', In A.J. Sbarra and R. Strauss (eds)

The Reticuloendothelial System, vol. 2 (Plenum Publishing Corp., New York 1980), p.355–68

Stephen, J. and Pietrowski, R.A. *Bacterial Toxins* (Nelson and Sons, Walton-on-Thames, England, 1981)

Taylor, P.W. 'Bactericidal and Bacteriolytic Activity of Serum Against Gram-negative Bacteria', *Microbiological Reviews* (1983), *47*, 46-83

Quie, P.G., Mills, E.L. and Holmes, B. 'Molecular Events During Phagocytosis by Human Neutrophils, *Progress in Hematology* (1977), *10*, 193–210

6 THE BACTERIAL CELL SURFACE IN RELATION TO THE ENVIRONMENT

The surface layer serves as the interface between bacteria and the outside world and as such plays an important role in the relationship between the organism and its habitat. In the natural environment extracellular polysaccharides constitute a notable feature of a majority of bacterial species. It is known that the exopolysaccharide endows the bacterial cell with a net negative charge and a hydrophilic surface. However it seems unlikely that the bacteria would commit such a large part of their metabolic energy to produce structures with such a limited passive role. The polysaccharide surface layers serve as the outermost mediators between the organism and the environment, the first point of entry and the last barrier to excretion. However, removal of capsular material, by modification of cultural conditions, enzymic degradation or mutation, has been shown not to affect the viability *in vitro* of a wide range of normally capsulated bacterial species. The significance of these observations should not be over-rated, since it is probable that the presence of the capsule only confers survival advantages under the highly competitive conditions of natural habitats. The diversity of structures possible in polysaccharide polymers far exceeds that of other macromolecules, making them invaluable as agents of specificity and diversity. Taking the simplest case, involving the linkage of two hexose sugars to produce a reducing disaccharide, eight possible combinations exist, depending on to which hydroxyl group the non-reducing sugar is attached and the configuration of the glycoside linkage. With the introduction of additional sugars the number of potential new combinations increases dramatically. A heterosaccharide chain containing four different monosaccharides can be arranged in more combinations than twenty amino acids in a tetrapeptide. The diversity of exopolysaccharides, and to a lesser extent, the polysaccharides of the O-side chains of LPS, are responsible for specificity in host-pathogen and host-symbiont interactions in plants. In addition bacterial polysaccharides can be implicated in the induction of disease symptoms and host defence reactions.

Bacterial polysaccharides have other roles in the natural environment that depend less on their antigenic specificity. Polysaccharide slime secreted by bacteria can serve as an adhesive, allowing microbial exosystems to resist the scouring action of water flow. The maintenance of good 'crumb' structure in soils depends upon the exopolysaccharides of bacterial origin.

Membranes are also important in the relationship between bacteria and the natural environment. The ability of bacteria to colonise extreme environments is accompanied by gross changes in membrane composition.

Plant-Bacteria Interactions

The relationships between bacteria and plants have important economic implications. The activities of the nitrogen-fixing symbiotic bacteria of root nodules are crucial to soil fertility. On the negative side bacterial pathogens are responsible for significant worldwide crop losses. The specificity of bacterial pathogens and symbionts for a particular plant host, the mechanisms involved in the induction of disease symptoms and the triggering of plant host defence mechanisms are all mediated by components of the bacterial cell surface.

Plants defend themselves from bacterial invasion by coating vulnerable surfaces with a protective layer. The aerial portions of the plant are covered by a cuticle composed of fatty acids and waxes secreted by the underlying epidermis. The intercellular spaces of the leaves are also protected by a thinner cuticle layer. Many plant tissues contain special secretory cells, producing poorly characterised materials, thought to provide a non-specific host-defence mechanism. Plant roots produce a mucilaginous layer, while the tip produces an acidic exopolysaccharide, both being implicated in protecting the underlying tissues from bacterial attack. Plants are subject to wounding by wind, browsing animals and contact with abrasive surfaces. A wound surface is essential for the establishment of many plant-bacteria interactions.

Pathogenic Interactions

The Gram-negative bacterium *Agrobacterium tumefaciens* induces crown gall tumours in many plants. The disease is especially important in parts of the United States of America, where if not controlled it can greatly reduce crop yields. *A. tumefaciens* can cause tumours in many dicotyledonous plants, but the disease has not yet been established in monocotyledons, e.g. grasses. The organism only produces symptoms when introduced into a pre-existing wound, apparently possessing low invasive potential. Plant cells from crown gall tumours can be grown in cell-culture and, unlike normal plant cells in such systems, they grow without added plant hormones. Once induced, the tumour cell will continue to grow in the absence of bacteria. The tumours, free of bacteria, when transplanted to healthy plants also continue to grow. Tumour formation is induced by part of a bacterial plasmid called Ti, which is transferred from *Agrobacterium* to plant cells. Strains of *A. tumefaciens* that have lost the Ti plasmid cannot cause tumours. Ti plasmids are large (M_r 9×10^7 to 1.6×10^8) and readily transfer between agrobacteria strains by conjugation. Only a small fragment (M_r 1.5 × 10⁷) of the Ti plasmid is maintained in the tumour cells. The Ti plasmid fragment causes changes in the concentration of certain hormones, which stimulate the abnormal growth of plant cells. In addition Ti plasmids induce biosynthesis by the plant of opines, compounds unique to crown gall tumours. Agrobacteria are able to utilise opines as sources of carbon and nitrogen.

Tumour-forming strains of *A. tumefaciens* have both chromosomal and plasmid-borne genes, each capable of specifying binding at distinct sites. Lipopolysaccharide (LPS) extracted from *Agrobacterium* is an effective inhibitor of tumour initiation if added before the infective bacteria, but not if added later. LPS extracted from non-binding strains of the organism is not inhibitory to tumour formation. The polysaccharide O-side chain appears to mediate adherence, since isolated O-antigen from *A. tumefaciens* is as active as the complete LPS and the lipid A moiety is non-inhibitory. Purified LPS from agrobacteria binds specifically to polygalacturonic acid, the LPS failing to bind to other plant carbohydrates or to methylated polygalacturonic acid containing plant tissue. Hence adherence, the primary event of infection by *Agrobacterium* strains bearing a Ti plasmid, is initiated by a specific interaction between two carbohydrates, the O-side chain of LPS, present at the surface of the bacterium, and polygalacturonic acid residues in the plant, possibly exposed by wounding.

Many plants exhibit a hypersensitive reaction in response to microbial invasion of the intercellular spaces. The reaction is rapid (6–10h) and consists of a localised desiccation and necrosis. Truly successful pathogens do not elicit this response and are able fully to parasitise the host. The hypersensitive reaction can be seen as the means by which the plant can limit the growth of the invading organism. Bacterial viability is essential to induce the hypersensitive reaction, since heat-killed *Pseudomonas solanacearum*, *P. lachrymans*, *P. tabaci* and *Xanthomonas axanapodis* do not give any response themselves yet impair the ability of viable cells of the same species to produce the hypersensitive reaction. Strains producing the reaction are said to be incompatible with the host and live as non-destructive saprophytes. Strains of the same species that fail to produce the reaction are compatible and can be considered as true pathogens.

Incompatible strains of *Pseudomonas pisi* appear to be immobilised in tobacco leaf tissue by adherence and this is followed by engulfment by cuticle-like material secreted by plant cells. The compatible tobacco pathogen *P. tobaci* does not produce such a response. Similarly the saprophytic bacterium *P. putida* adheres to and becomes enveloped by cells in the bean leaf, while a compatible pathogen *P. phaseolicola* shows no response. Actual physical contact between the bacterium and plant cell walls is required for the hypersensitive response in pepper plants induced by *X. vesicatoria*.

The presence of an extracellular polysaccharide slime is a pre-requisite in virulent strains of *P. solanacearum* for tobacco and potato plants: loss of slime production renders the strain avirulent. Many plants produce proteins that can agglutinate mammalian red blood cells, sometimes called phytohaemagglutinins, but better known as lectins. Virulent and avirulent isolates of *P. solanacearum* may be distinguished by reaction with a lectin isolated from potato. Virulent strains possessing an extracellular poly-saccharide are not agglutinated by potato lectin. In contrast, the non-pathogenic capsular strains are readily agglutinated by the lectin and produce a hypersensitive response. Lectin formation can be considered as a host defence mechanism, the lectin reacting with non-encapsulated bacteria, fixing them to plant cells, producing the hypersensitive reaction, thereby limiting the infection. Lipopolysaccharide (LPS) from both smooth and rough strains (i.e. with and without the O-side chain) inhibit the hypersensitive response, indicating that the polysaccharide of the side chain has little importance in the reaction. Acetic acid hydrolysis and alkaline deacylation of the LPS destroyed this activity, suggesting that both the fatty acids of the lipid A and the core-lipid A linkages are necessary for activity. The polysaccharide at the surface of virulent strains masks the LPS, preventing lectin binding, and the hypersensitive response, allowing the organism fully to parasitise the host. This mechanism is not unique to *Pseudomonas* similar observations have been recorded in *Erwinia amylovora* in apple, and *Xanthomonas oryzae* in rice.

Xanthomonas produces an exopolysaccharide termed xanthan, which is responsible for highly specific bacteria-plant interactions. Xanthans exist as helical polymers of mannose and galactose. Purified xanthan associates strongly with glucomannans and soluble celluloses. In nature, xanthan secreted by the bacterium is capable of highly specific interaction with plant cell-wall polysaccharide polymers to produce a stable gel. This interaction serves as a strong adhesive capable of anchoring the pathogen to its host and initiating the pathogenic process.

Symbiotic Interactions

Bacteria of the genus *Rhizobium* engage in a symbiotic relationship with leguminous plants. In return for organic nutrients and shelter, the bacterium provides its host with nitrate. Nitrogen fixation is not restricted to *Rhizobium*; many other bacteria and certain blue-green algae are also capable of converting atmospheric nitrogen into nitrate, but only rhizobial nitrogen fixation is used in current agricultural practices.

It is a characteristic of the rhizobium-legume symbiosis that the bacteria exhibit great specificity in their choice of host, indeed the genus *Rhizobium* is divided into species on the ability of the strain to form nodules in specific plant species (Table 6.1). Inter-species and inter-strain nodulation may occur, although often the nodules produced are atypical and fail to fix nitrogen effectively.

Table 6.1: Host Distribution of Nitrogen Fixing Root Nodule Bacteria of the Genus *Rhizobium*. In addition certain strains, not listed have a wider spectrum of activity, e.g. the 'promiscuous' strains effecting the cow pea groups.

Rhizobium species	Plant host capable of forming effective nodules
R. trifolii	Clover (*Trifolium*)
R. leguminosarum	Pea family (*Lathyrus, Vicia, Pisum, Lens*)
R. phaseoli	Bean (Phaeseolus)
R. lupini	Lupins (*Lupinus, Ornithopus*)
R. japonicum	Soybean (*Glycine*)
R. meliloti	Alfalfa (*Medicago, Trigonella, Melitolus*)

Colonisation by rhizobium begins shortly after uninfected roots grow into soil containing free-living infective phase *Rhizobium* cells. The bacteria show a chemotactic response to metabolites diffusing from the legume root and become aligned at right angles to a root hair cell. These hair cells are protected by a mucilaginous material, and the *Rhizobium* cells become embedded in this layer (Figure 6.1). Lectins produced by the host bind to both the root hair surface and to the surface of a compatible strain of *Rhizobium*, i.e. it is the binding specificity of the lectin produced by the plant host that determines which *Rhizobium* strains can form a nodule. Microscopic examination of legume roots *in situ* shows that the attachment of the bacterium to the root hair is restricted to the polar region. After the initial attachment by lectins is complete, the bacterium lays down cellulose fibrils which cement the union. The bacterium then breaches the cell wall of the hair cell and forms an infection thread, which progresses by a growing invagination of the host cytoplasmic membrane. Infected root hairs exhibit a characteristic curling. The bacteria enlarge, undergo division, and ultimately induce the formation of the nitrogen-fixing nodules (Figure 6.2). As the nodules age they decay and release free bacteria back into the soil.

The economic importance of nitrogen-fixation by leguminous plants has led to extensive study of the process of nodulation and in particular the nature of the host specificity of *Rhizobium*. Antiserum raised to clover tissue reacts with virulent strains of *R. trifolii* and to a lesser extent with avirulent strains of the same organism. If the antiserum is pretreated with avirulent cells a purified antibody, specific for the virulent strain, can be prepared. The purified antisera will bind to *R. trifolii* virulent strains and clover cells, but not to incompatible *Rhizobium* species. Immunofluorescence can be used to demonstrate that antisera raised against the capsular polysaccharide of *R. trifolii* binds to the surface of clover root cells, in particular to the tips of the root hairs. Chemical analysis of this capsule shows that it

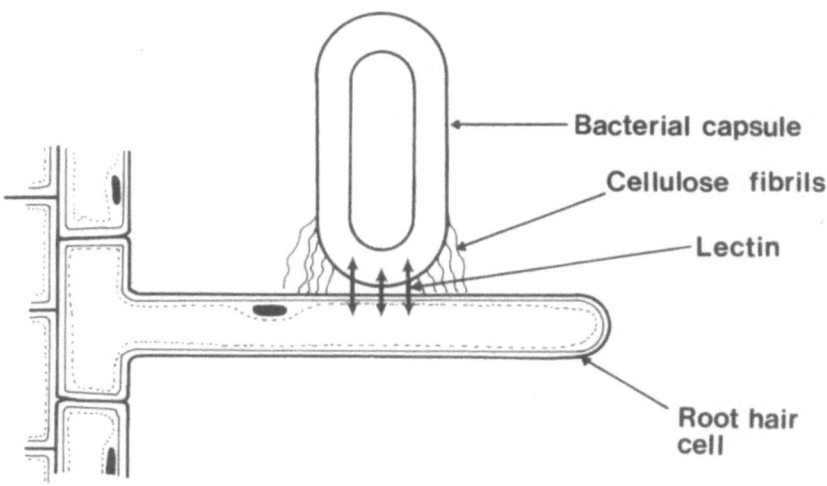

Figure 6.1: Mechanism of Binding by *Rhizobium* to Root Hairs of Leguminous Plants. In reality the root hair is not straight but highly convoluted and much longer than shown. The bacteria bind predominantly to the root tip, which is thought to possess elevated levels of lectin.

consists of a polysaccharide polymer, containing β-linked 2-deoxyglucose, galactose and glucuronic acid. Purified capsule alone can induce curling of the root hairs. Trifoliin, a lectin isolated from clover, specifically agglutinates infectious, but not non-infectious mutants, of *R. trifolii* or other infective *Rhizobium* species. The agglutination can be inhibited by 2-deoxyglucose and *N*-acetylgalactosamine, suggesting that the former, which is present in the bacterial exopolysaccharide, may be the site of lectin binding. In the presence of 2-deoxyglucose, but not other sugars, the binding of *R. trifolii* to clover roots is inhibited. The adherence of other *Rhizobium* species to their hosts is not inhibited by 2-deoxyglucose, suggesting that in other species, different sugars may be involved in recognition and adherence. Trifoliin has been shown to be present in greatest amount at the root hair tip and is eluted from the clover root by 2-deoxyglucose.

Isolated LPS also binds to lectins, although this may be less important in the recognition process. LPS isolated from *Rhizobium* species only binds to lectin derived from its natural plant host. The polysaccharide of LPS O-side chain is responsible for lectin-binding. In most *Rhizobium* species only the restoration of exopolysaccharide product can confer infectivity on

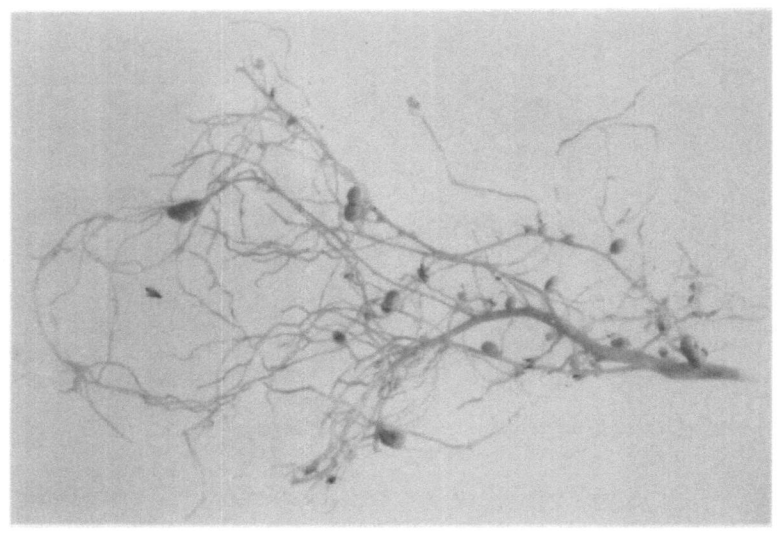

Figure 6.2: The Effect of the Possession of Nitrogen-fixing Root Nodules on the Growth of Clover. The plant on the left was grown in sterilised soil, the one on the right in soil containing *Rhizobium trifolii*. Roots taken from the plant on the right show extensive nodulation.

avirulent mutants, even though these mutants bear complete O-side chains. There is, however, one notable exception, soybean lectins fail to react with the extracellular polysaccharides of *R. japonicum* but strongly react with O-antigenic polysaccharides, suggesting that LPS may be involved in host recognition in this species.

Fimbriae of Saprophytic Bacteria

Many free-living saprophytes, such as *Klebsiella aerogenes*, *Enterobacter cloacea* and *Serratia marcescens*, possess type 1 mannose sensitive fimbriae (p. 140). Saprophytes may benefit from the ability to adhere to plant and fungal tissue capable of yielding organic nutrients, from which they might otherwise be removed by, for example, rain water flow. Saprophytes inhabiting stagnant anaerobic water, utilise fimbriae to adhere to the pellicle forming at the air/water interface, giving the organisms access to oxygen diffusing from above.

The Role of Bacterial Surface in Freshwater and Marine Environments

Many of the heterotrophic bacteria present in fresh or sea water exhibit a tendency to adhere to surfaces, often followed by the production of large amounts of extracellular polysaccharide. Attached bacterial populations are of greatest importance in low nutrient waters, as here numbers of suspended bacteria are small. In aqueous bacteria there is little evidence of specific attachment organelles, such as pili. However, certain members of the genera *Caulobacter* and *Asticacaulis*, develop holdfasts at the flagellar pole of the cell, which enable the bacterium to attach to surfaces, as well as to some other microorganisms. The majority of aqueous bacteria possess a thin hydrophilic, negatively charged surface coating in the form of a polysaccharide capsule of slime, which mediates adherence.

Freshwater Bacteria

Examination of stones on the bottom of fast flowing unpolluted streams shows them to be covered in a thin slime layer. Bacteria can be readily isolated from the slime by rubbing the stones over the surface of bacteriological media. Such isolates are invariably Gram-negative species, and many of these bacteria can produce their own weight in polysaccharide in 24 h. Analysis of the exopolysaccharide of aqueous bacteria shows that they often contain D-glucose, D-mannose and D-galactose and less commonly L-rhamnose. In addition they contain uronic acids, either D-galacturonic or D-glucuronic acids. The polysaccharides vary in the degree of hydration, some being highly viscous and watery, others being less viscous with a low water content. The polysaccharide of freshwater strains of *Serratia marcescens* have been analysed in some detail, and shown to consist of D-mannose, D-galactose and D-glucuronic acid. Interestingly, polysaccharides from freshwater isolates do not contain *O*-acetyl groups, whereas such substituents are common in polysaccharides of non-aquatic bacteria.

Rivers receiving persistent discharges of organic pollutants, contain large numbers of the Gram-negative chain-forming rod, *Sphaerotilus natans*. Under the conditions prevalent in a river containing high levels of organic matter, the individual bacteria are embedded in long filamentous sheaths, made up of protein, polysaccharide and lipid. The bacterial filaments add to one another, forming a fluffy mass known as 'sewage fungus'. The filaments elongate by division of the cells, within the sheath. Certain cells may acquire flagella break out of the filaments and thereby spread the infection. In addition this organism rapidly coats solid objects on the river bed with a thick polysaccharide layer. The carbohydrate of the exopolysaccharide is distinct from the protein-polysaccharide-lipid complex that forms the sheath. The *Sphaerotilus* acidic exopolysaccharide is chemically related to colanic acid produced by *E. coli*, in that it contains approximately equal amounts of fucose, glucose, galactose and glucuronic acid. In addition the polysaccharide contains poly-β-hydroxybutyrate and glycogen- Riverine bacteria can produce vast amounts of polysaccharide, if sufficient nutrient is present. It is estimated that up to five tons of bacterial slime pass a given point on the banks of the Danube each day. In the nutrient rich waste effluents discharged by the food and dairy industry and in pulp paper manufacture, large amounts of slime develop. Effluents with high sucrose contents, especially encourage dextran- and levan-producing organisms. The production of slime in rivers is not only deleterious to the aesthetic quality of a river. The slime can move downstream, where its breakdown by other microorganisms causes local deoxygenation, resulting in asphyxiation of fish and other animals.

Little information is available concerning the production of microbial polysaccharides in standing bodies of freshwater. Heterotrophic lake bacteria grow slowly, their biosynthetic abilities depending upon the levels of nutrient availability. At the lake surface, phytoplanktonic algae predominate. The decay of the algal population provides the carbon source for the heterotrophic bacteria, which are usually present in natural waters in direct proportion to the levels of organic matter. The living community that develops, during calm weather, on the surface of a lake, is often held together by the formation of a slime layer by *Pseudomonas*, *Caulobacter* and Flavobacteria.

In rivers containing large numbers of bacteria, parasite vibrios of the genus *Bdellovibrio* are common. These monoflagellated vibrios attach to Gram-negative bacteria, penetrate the envelope and internally parasitise the host. Polysaccharide capsules appear to provide little protection to the host. However rough mutants of *Salmonellae typhimurium* and *E. coli* (i.e.

lacking the O-side chain but containing the core region of the lipopoly-saccharide) are more sensitive to attack by *Bdellovibrio* than the cor-responding smooth strain (i.e. with the O-side chain).

In grossly polluted rivers supporting large bacterial populations a wide variety of bacteriophages are present and play an important role in control-ling bacterial numbers. In order successfully to parasitise their host, bac-teriophages must first adhere to the bacterial host. Certain of the phages of enteric bacteria exhibit binding to specific O-side chain configuration as well as their core structures or to capsular polysaccharides. In addition phages such as T1, BF23 and λ react with proteins in the outer membrane. However the approach of core-specific phages to their receptors is hindered by the presence of an O-side chain on the LPS, despite the fact that other phages penetrate through the LPS to reach protein receptors in the outer membrane. In Gram-positive bacteria, the presence of extracellular polysaccharide may hinder the approach of phage to surface receptors.

Marine Bacteria

The adherence of marine bacteria to surfaces has received much attention, since colonisation by bacteria is the first stage of the development of a microbial ecosystem on immersed objects. Surface-colonising types are commonly referred to as periphytic microorganisms. This phenomenon is of great economic importance as it precedes 'fouling' of ships bottoms and man-made structures, e.g. oil rigs. A ship bearing considerable amounts of fouling experiences viscous drag, slowing the vessel and increasing fuel costs. In addition adhered bacteria can provide foci for corrosion of metal components; *Desulphovibrio* and *Desulphotomaculum* being responsible for microbial corrosion of ferrous metal, *Leptothrix* and *Bacillus mycoides* for aluminium plates. Bacterial attachment also has implications in the production of commercial food crops, e.g. oysters. When objects are immersed in the sea they are rapidly colonised by successive populations of microorganisms. Within a few hours of placing a clean smooth surface, e.g. a sterile glass plate, in the sea many marine chemo-organotrophs, prim-arily Gram-negative rods, *Pseudomonas, Flavobacterium* and *Achromo-bacterum*) can be found adhering to the surface. Within two to three days, the microflora changes and the predominant colonisers are species of *Caulobacter, Hyphomicrobium* and *Saprospira.* One week after sub-mersion the microbial ecosystem is fully developed, the total bacterial population reaching about 2×10^6 bacteria cm^{-2}. The bacterial population exists within a secreted polysaccharide film which coats the substrate. All periphytic marine bacteria, so far described, secrete a non-diffusible uronic acid-containing polysaccharide. The development of the bacterial film is followed by further colonisation by attaching protozoans, diatoms, the floating planktonic larvae of barnacles (Cirripedia), tube worms

(Polychaetae), bryozans (Polyzoa) and seaweeds (Fucoidea).

Extracellular polysaccharides produced by the bacteria play a vital role in the adherence of marine bacteria to surfaces. It is possible to recognise two distinct adhesive mechanisms. On the surface of both attached and free-living bacteria can be detected a thin polysaccharide coat believed to be responsible for the initial interaction between the bacterium and substrate. Spontaneous attachment occurs when a bacterium encounters a substratum and comes close enough for the polymeric coat to become adsorbed. The polysaccharide bears a net negative charge, but if the presented surface is neutral or positive (or if the mutual electrostatic repulsive forces are not prohibitively large) the bacterium can adhere to the surface. The degree of hydrophobicity of the underlying substratum has an effect upon the attachment of marine pseudomonads to the surface. Hydrophobic substrates, e.g. polystyrene, become completely colonised within a few hours and ultimately become coated with large numbers of bacteria. Adherence to hydrophilic surfaces occurs much more slowly, the number of organisms found on negatively charged surfaces, e.g. glass or mica, being small. However, if the glass surface is made more hydrophobic, by treatment with trimethylchlorosilane, the number of colonising species and the total number of bacteria found adhering to the glass dramatically increases. If some degree of hydrophobic interaction were not favoured there could be a tendency for the organisms to remain in suspension, rather than to adhere to a hydrophilic substratum.

After the initial attachment the adhered cells multiply and are joined by additional cells, leading to the formation of micro-colonies. Copious amounts of extracellular polymers may also be produced; electron microscopy demonstrates that the bacteria within the slime layer are embedded in a polymeric matrix, which is largely polysaccharide. Whether the primary polysaccharide responsible for the initial interaction and the secondary polymer responsible for the formation of the slime layer are chemically related is uncertain. However, if a test plate is transferred to cation-deficient media, the secondary polymer is disrupted suggesting that Ca^{2+} and Mg^{2+} are important in maintaining the adhesive polysaccharide structure. It is unclear whether bacterial adhesive polymers undergo a setting process to become permanent adhesives, although electron micrographs of intracellular matrices give some indication that with time the polymer loses some of its original elasticity, possibly through cross-linkage reactions. If the polysaccharide contains more than one polymer type, co-gellation may occur, and it is even possible that, where more than one bacterial species is present in the matrix, some form of symbiotic association of the polysaccharide is necessary. The completed polymer exists as a highly hydrated gel, forming an intercellular matrix. The adhesive polymers of most marine slime formers are readily stained with ruthenium red/ osmium tetroxide, a procedure thought specific for acidic polysaccharides.

Chemical analysis of the adhesive polymers of marine pseudomonads shows that in addition to the polysaccharide there is a significant protein component. The carbohydrate contains mannose, glucose, glucosamine, rhamnose, galactose and ribose. Since infra-red spectroscopy and histochemical techniques suggest the presence of free carboxyl groups and chemical analysis fails to detect uronic acids, the acidic groups may in fact be located in the protein fraction. The large number of carboxyl groups present in the extracellular polymers may require the presence of divalent cations for screening of charged groups of certain cross-linkage interactions, essential for the maintenance of the structural integrity of the adhesive. An exception is the polymer of *Pseudomonas atlantica*, which contains mannose, glucose and galactose together with galacturonic acid and pyruvate. This polymer differs from others in that the negative charge is produced by high levels of uronic acids (ratio of hexose to uronic acid approximately 1:1). Further negative charge is contributed by a pyruvate content of approximately 10 per cent.

Bacteria isolated from marine slime films, possess a wide range of proteolytic and polysaccharide-degrading hydrolytic enzymes. *Vibrio parahaemolyticus* absorbs specifically to chitin particles and plankton present in the marine film. The organisms possess enzymes capable of degrading and utilising the chitin components of the film.

Role of Adherence in the Aqueous Ecosystem

There is much debate concerning the possible advantages which could be derived by the bacterium by attaching to a surface, rather than living exclusively in the aqueous phase. Attachment may encourage microbial growth by providing a micro-environment which is rich in adsorbed macromolecular nutrients, as compared to the surrounding aqueous environment. The interstices of the gel may serve as foci for the concentration of extracellular enzymes necessary for the degradation and subsequent assimilation of macromolecular nutrients. A thick slime layer may also protect the bacterial ecosystem within from deleterious changes in the external world. Surface bacterial cells may detach from the fully developed polysaccharide polymer. Thus the slime layer acts as a bacterial reservoir, capable of repopulating the aquatic environment, when conditions become more favourable. The matrix also provides some protection from the abrasive action of particulate matter carried in freshwater streams and by the scouring action in the intertidal zone of the sea. A price must be paid for living within the polysaccharide slime layer. Growth within the matrix is often limited by the slow diffusion of nutrients and metabolic products or oxygen through the polymer.

Bacterial Exopolysaccharides in Sewage Treatment Plants

In the United Kingdom rainfall, household sewage and industrial wastes find their way into sewers. The sewage contains organic and inorganic substances, both in solution and suspension. Organic compounds constitute approximately 90 per cent of the total soluble carbon content, containing carbohydrates (29 per cent), amino acids (11 per cent), organic acids (40 per cent), creatinine (4 per cent) and anionic surfactant detergents (15 per cent); however the addition of quantities of industrial effluents may drastically alter this. A medium size works serving a population of 20,000–30,000 will deal with about 1 million gallons of sewage/day. By law, before this can be discharged into a British river its biologically oxidisable organic content, known as the Biochemical Oxygen Demand (BOD) must be reduced to 20 mg l^{-1}. The BOD of raw sewage may exceed 400 mg l^{-1}. In the first stage of any treatment plant as much insoluble matter as possible is precipitated as a liquid sludge, by slowly passing the sewage through sedimentation tanks. This purely physical process removes approximately half the BOD. The only practicable way of removing the remaining soluble organic matter from large volumes of settled sewage (i.e. raw sewage after primary sedimentation) at reasonable cost, is by the action of bacteria under aerobic conditions. To ensure complete oxidation of the organic compounds present, the sewage and oxygen must be brought in contact with large masses of active bacteria. The bacterial mass will constantly increase (a medium size works will produce up to 5 tons of bacteria/day), and this must be removed at a rate equivalent to the rate of microbial growth. Two distinct biological oxidation processes for the treatment of settled sewage are used in the UK. Extracellular polysaccharides play an important role in both processes.

Biological Filtration. The settled sewage is passed down through a percolating filter. This usually consists of a bed of filtering matrix, i.e. stones, clinker or more recently plastic, made up of particles approximately 2 inches in diameter. The beds are circular or rectangular and are usually about 6 feet deep (Figure 6.3). The settled sewage is evenly distributed over the bed surface by rotating arms or in rectangular beds by a distributor moving backwards and forwards. The filtering matrix rests on a concrete floor, provided with underdrains and ventilating tiles to provide aeration by natural draught. The surface of the filter bed medium becomes coated with a film of bacteria and as the settled sewage passes over the medium it comes into contact with the large numbers of bacteria embedded in the film. The optimum film depth required to produce a high quality effluent is about 0.15 mm.

The bacterial population of the filter bed is adapted to many different substrates and the majority of organic wastes can be oxidised. Bacteria associated with the different stages of sewage breakdown are established at distinct levels in the bed; heterotrophs at the surface and autotrophs nearer the base of the bed (Figure 6.4). Species able to colonise the bed find it an

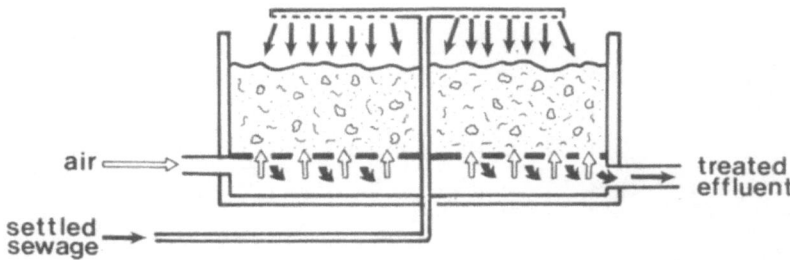

Figure 6.3: Biological Filter Bed. Settled sewage is distributed over the surface of the bed containing stones, clinker or plastics. The sewage percolates downwards, where it is oxidised by bacteria held in a thin polysaccharide film covering the bed matrix. A natural air draught upwards through the bed provides sufficient oxygen to metabolise soluble organic material in the sewage.

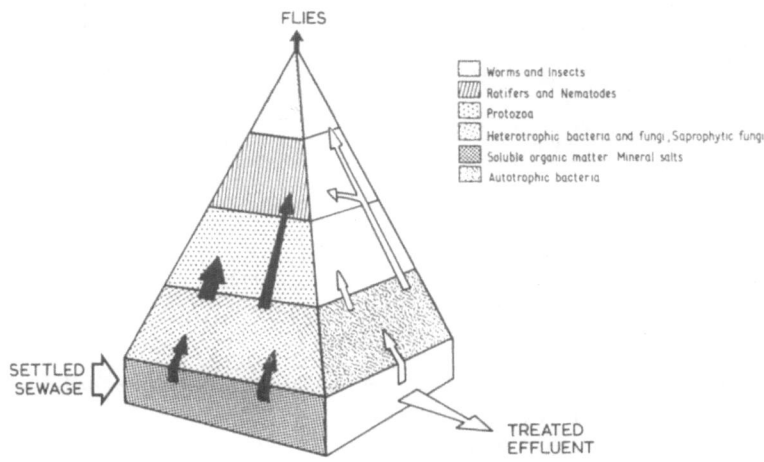

Figure 6.4: Energy Flow Through the Filter Bed Ecosystem. Bacteria and fungi present in the zoogleal film assimilate nutrients from the settled sewage and constitute the first tropic level. Bacteria and fungi are eaten by the second trophic level mainly protozoans. Rotifers and nematode predators constitute the third trophic level. At the summit of the food pyramid are worms and insect larvae. The bed represents a complex and highly integrated ecosystem in which the bed community is responsible for complete oxidation of the organic content of the sewage.

environment well suited to rapid bacterial growth. Sewage is generally above ambient temperature and contains an abundance of nutrients.

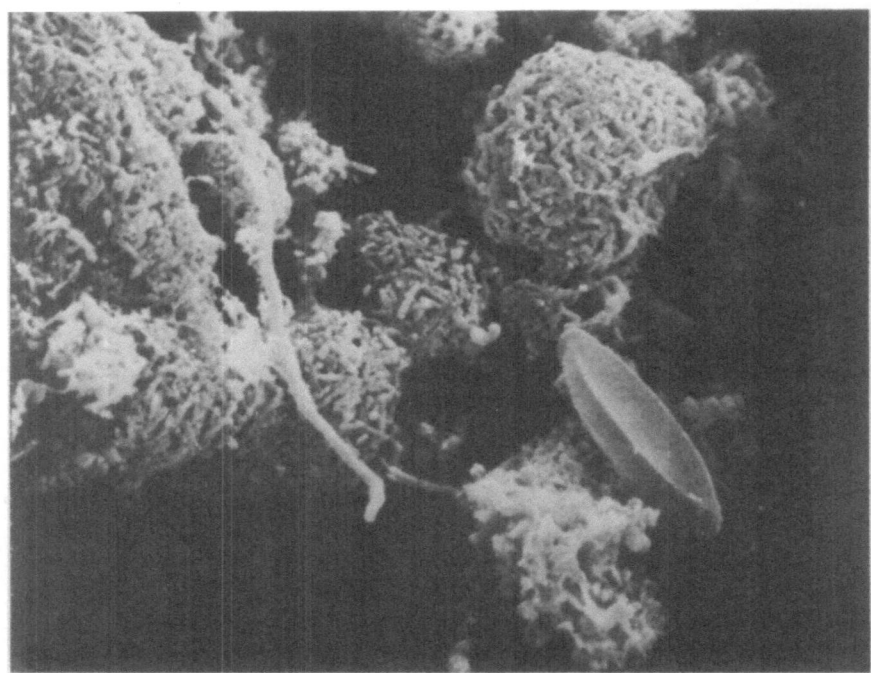

Figure 6.5: Scanning Electron Micrograph of a Mixed Bacterial Film Growing on a Filter Bed. The large elliptical object is a diatom. (Photograph by permission of Mack, W.N., Mack, J.P. and Ackerson, A.O. (1975) *Microbial Ecology,* 2, 215.)

However this environment is rather specialised and natural selection tends to restrict the number of species present.

In an active filter bed the supporting matrix is coated with a gelatinous polysaccharide film secreted by the bed bacteria. The species involved varies from plant to plant, depending upon the discharges received, but they are invariably Gram-negative, aerobic, non-sporing, encapsulated rods (Figure 6.5). *Zoogloea ramigera*, a slime-forming pseudomonad, is probably the most common film bacteria. The organism appears to be distributed randomly in a polysaccharide gel, from which long filaments containing bacteria, may extend. Within the slime is found quantities of poly-β-hydroxybutyrate and fibres of cellulose. The purified matrix polysaccharide isolated from *Zoogloea* absorbs metal ions from solution. This may serve as a mechanism conferring resistance to the heavy metals often found in industrial effluents. Other film-forming bacteria, including

Sphaerotilus natans and the sulphur bacterium *Beggiatoa alba*, may also coat the solid surfaces of the bed. Non-film forming bacteria, e.g. *Alcaligenes*, achromobacteria, flavobacteria and pseudomonads become embedded in the zoogleal film. Organic solutes and dissolved gases can freely permeate the film. The predominant chemotrophs in the bed are the nitrifying bacteria, i.e. those bacteria responsible for the conversion of ammonia (from urea and protein) into nitrite and ultimately nitrate.

$$NH_4^+ \xrightarrow[\text{\textit{Nitrosomonas}}]{O_2} NO_2^- \xrightarrow[\text{\textit{Nitrobacter}}]{O_2} NO_3^-$$

Nitrosomonas and *Nitrobacter* are present through the entire depth of the bed, however as one would expect the proportion of *Nitrobacter* increases with depth in response to the production of nitrite by *Nitrosomonas*.

If the film bacteria were allowed to grow unchecked, the film would begin to fill the interstices between the bed particles, the bed would become clogged, passage of fluid would halt, and the bed would become anaerobic, septic and non-functioning. This is prevented by the large numbers of holozoic protozoa, e.g. *Carchesium, Paramecium, Amoeba, Epistylis* and *Stylonychia* which feed upon both the bacteria and the zoogleal film. In their turn the protozoans are controlled by a grazing fauna which includes Rotifers, insect larvae and worms.

Activated Sludge Process. In this process the settled sewage is held in aerated tanks. Although differing in detail, the basic principle of the many plants using this process is to get enough oxygen into solution to allow bacteria to oxidise all the organic waste. Before the settled sewage enters the aeration tank it is seeded with actively growing bacterial sludge. This provides the large numbers of bacteria necessary for the rapid oxidation of the soluble organic fraction. Settled sewage is normally completely degraded in 4–8 h. The dissolved oxygen level of the tank must be continuously maintained at about 1 mg oxygen/l, to permit complete conversion of carbohydrate into carbon dioxide and water, and ammonia into nitrate. The bacterial biomass formed during the activated sludge process, both dead and alive, their decomposition products and unchanged suspended solids have a tendency to form aggregates termed 'flocs' and can be removed by subsequent sedimentation. After passage through the aeration tanks, when all the organic matter has been oxidised, or converted into bacterial biomass, the treated sewage enters a secondary sedimentation tank. The bacterial flocs fall to the base of the tank, where they collect, and the treated sewage leaves over a peripheral weir. About 6 per cent of the collected bacterial sludge from this tank (the 'activated'

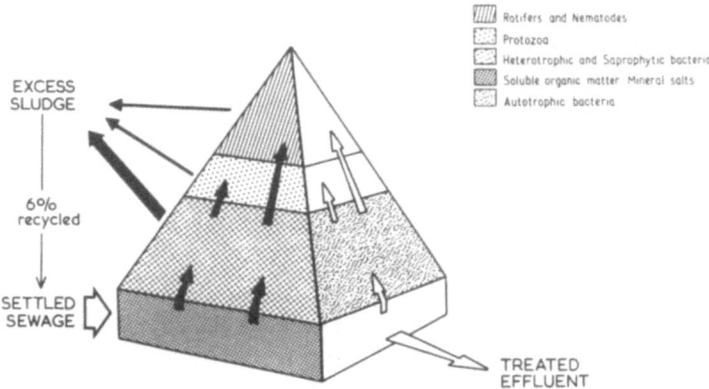

Rotifers and Nematodes
Protozoa
Heterotrophic and Saprophytic bacteria
Soluble organic matter Mineral salts
Autotrophic bacteria

Figure 6.6: Energy Flow Through the 'Activated Sludge' Process. The ecosystem is less complex than that of the filter bed (Figure 6.4). There are fewer trophic levels and the protozoa play only a minor role in purification. The bulk of the organic matter of the settled sewage is converted into bacterial biomass which is aggregated into polysaccharide based 'flocs', removed by sedimentation, i.e. most of the available energy does not pass along the food chain, but is removed and dumped.

sludge) is returned to re-inoculate the aeration tank, the remainder is dumped.

One could consider the flora and fauna of the filter bed to be a direct development from the populations found on mud banks of polluted streams, indeed both ecosystems share many common species. However, activated sludge is basically an aquatic environment and contains a more restricted range of organisms than the filter bed (Figure 6.6). Under optimum conditions, the bacterial population of the activated sludge is maintained at about 6 g/l, about 10–25 per cent by weight living bacteria. As in the filter bed, the bacteria constitute the primary trophic level. The majority of bacterial genera described from activated sludge are Gram-negative rods. *Zoogloea ramigera* and *Sphaerotilus natans* secrete a gel-like mass, which may serve as the basis of the 'flocs'. Once the floc has begun to form certain of the bacteria die and lyse. The insoluble poly-saccharides remain and entrap less-active bacteria. The bacteria entrapped in the flocs include achromobacteria, flavobacteria and pseudomonads.

The production of extracellular polysaccharides by *Sphaerotilus natans* and *Zoogloea ramigera* is vital to flocculation and without it the separation of the oxidising bacterial population would not be practicable. Under certain conditions the filamentous form of *Sphaerotilus natans* may become the major organism in the sludge. When this occurs the ability of the sludge to settle becomes impaired, resulting in an inferior final effluent, one which contains high bacterial numbers. The inability of oxidised sewage to settle is termed 'bulking' and may result from changes in oxygen levels, fluctuations in sewage strength or the discharge of toxic wastes. Changes in microbial extracellular polysaccharides are implicated in bulking of activated sludges.

Bacterial Polysaccharides in Soil

Polysaccharides are an important constituent of many soils. Hemicelluloses and celluloses, derived from plants, represent a significant source of energy and nutrients to soil micro-organisms. When plant residues reach the soil the hemicellulose degrades initially at a high rate, but other plant polymers, notably some celluloses, are more refractory to degradation. Since plant polymers cannot freely penetrate the bacterial envelope, because of their high molecular weight they must be first broken into simpler sugars by extracellular enzymes. A secondary microbial population also develops, capable of utilising compounds liberated from plant polysaccharides by the hemicellulolytic and cellulytic flora. The breakdown of the xylose-containing hemicellulose xylans have received great attention, since these polysaccharides make up the greatest proportion of the total carbohydrate content of woody plants and grasses. The soil contains a large number of fungi, bacteria and actinomycetes capable of xylan hydrolysis. In acid soils, the preliminary stages of xylan decomposition are accomplished by filamentous fungi, but in neutral or alkaline conditions bacteria, notably of the genus *Bacillus* but also *Pseudomonas* and *Erwinia*, predominate. *Bacillus* species are also important in the rotting of straw in manure, and in mature heaps the temperature rises as the result of microbial activity. At temperatures of 60 to 65°C, the xylan is rapidly hydrolysed by thermophilic aerobic spore-forming bacilli.

Cellulose is more refractory than hemicelluloses to bacterial attack. The numbers of aerobic, mesophilic bacteria metabolising cellulose depends upon soil composition and cellulose input, but may reach 10^6/gram of soil in manured fields and close to plant roots. Bacteria of the genus *Bacillus*, *Cellulomonas*, *Clostridium*, *Corynebacterium*, *Pseudomonas*, *Vibrio* and some others are capable of digesting cellulose. The most common anaerobic cellulose ferementers in nature are members of the genus *Clostridium*, present in soil, compost, manure and sewage. Many clostridia are cellulytic. The thermophile *Clostridium thermocellum* is capable of rapid hydrolysis of cellulose at 60°C.

Much of the degraded plant polysaccharide is converted into bacterial biomass. Radiolabelled [^{14}C]glucose, dextran and plant tissues have been added to soils to determine the fate of plant polysaccharides. Appreciable amounts of radiolabel become localised in bacterial exopolysaccharides.

Indeed examination of many soil bacteria *in situ* reveals them to be heavily encapsulated. The synthesis of bacterial polysaccharides appears crucial to the maintenance of soil structure and ultimately soil fertility. The coalescence of mineral particles to form aggregates and the resistance of such aggregates to stress is a measure of a soil's stability. Many soil bacteria are able to bind soil particles into aggregates, through the production of extracellular polysaccharides, or by charge interactions. Addition of *Azotobacter chroococcum* or *Pseudomonas* sp. promotes the stabilisation of sterilised silt loam soils.

Soil bacteria are subject to predation by phagotrophic protozoans and slime moulds. There is limited evidence suggesting that extracellular polysaccharides protect soil bacteria from predation. Slime-forming strains of *Pseudomonas fluorescens* are not ingested by *Amoeba limax*, whereas noncapsulated strains are readily phagocytosed. Under conditions of sucrose limitation this pseudomonad cannot synthesise extracellular polysaccharide and otherwise resistant strains are readily engulfed. Slime moulds (myxomycetes) grow on solid surfaces and obtain their nutrients by solubilising particulate insoluble substrates, especially bacteria, which they kill and lyse. The lytic activity of myxomycetes is mediated by an enzyme very similar to lysozyme. The enzyme cannot degrade the peptidoglycan of Gram-negative bacteria, unless the organism is presensitised. Fatty acids activate the bacteriolytic enzymes of myxomycetes and permit the lysis of otherwise resistant bacteria. When Gram-negative bacteria are digested by the myxomycete *Dictyostelium*, the lipopolysaccharides of the bacteria are degraded, with the release of the fatty acid component of lipid A. The freed fatty acid is transferred to the slime mould becoming available to serve in the next round of bacterial lysis. Species of *Rhizobium* are generally resistant to predation and this has been attributed to the unusual composition of their exopolysaccharides, which possess a high degree of sugar heterogeneity and often certain methylated sugars.

Bacterial growth in soil is limited by the availability of water. Vegetative soil bacteria are subject to periods of drought during which there is great danger of desiccation. Their survival depends upon a complex interplay of factors, one of the most important of which is believed to be the possession of extracellular polysaccharide. The hydrated polysaccharide gel provides a localised environment, possessing sufficient residual moisture to permit the survival of vegetative bacteria. In addition, many bacterial exopolysaccharides are hygroscopic and may serve to sequester water from moist air.

Bacteria in Extreme Environments

There exists in nature a series of environments which would, on the face of it, prove very hostile to living organisms. One of the great strengths of the bacterial kingdom is its adaptability; bacteria can be isolated under conditions that would kill most other life forms. Bacterial life can exist in hot springs some of which can also be very acid; lakes with a high salt content; very acid streams, such as found in mine effluents; in concentrated industrial acids, and alkalis; in the presence of high concentrations of toxic metals; on dry rock desert surfaces; in the ocean depths at pressures in excess of 1000 atmospheres. In hostile environments, the bacterial cell surface is important in protecting vulnerable cytoplasmic components from the adverse conditions without.

Growth at Low Temperatures

A large proportion of the Earth's biosphere is permanently cold. The polar regions represent about 15 per cent of the Earth's surface and 90 per cent of the oceans are below 5°C. Study of the organisms that can grow at temperatures below zero has important economic considerations, since considerable amounts of foodstuffs are now preserved by freezing. Organisms capable of growing at low temperatures are called psychrophiles.

Study of electron micrographs of most psychrophiles does not suggest that specialised organelles or surface layers play a role in adaptation to low temperature. However, if the organism is grown at room temperature, gross changes in cellular morphology are observed. At temperatures above the normal growth range, the cell walls of *Vibrio psychroerythus* and *Bacillus psychrophilus* are broken down, suggesting that these envelope layers are temperature sensitive.

The lipid composition of membranes is a sensitive indicator of changes in environmental temperature. The fluidity of a membrane is critical to its functioning as a semi-permeable barrier, and is directly related to the fatty acid composition of the membrane. In artificial lipid membrane the liquid to crystalline transition occurs at lower temperatures for phospholipids containing higher proportions of shorter chain fatty acids or increased degree of unsaturation. Bacterial membranes with a greater proportion of unsaturated fatty acids are better able to function at low temperatures.

Growth at Elevated Temperatures

Obligate thermophiles demonstrate optimum growth at temperatures above 65°C. *Bacillus stearothermophilus*, which grows well at between 65 and 70°C has received much study, since it can survive inadequate canning or bottling, producing subsequent food spoilage. However, *B. stearothermophilus* is only a moderate thermophile. Recent attention has turned to *Thermus aquaticus*, isolated from hot springs in Yellowstone National Park, USA, which grows best at 80–85°C. This temperature, at the elevation of Yellowstone, is only just below the boiling point of water.

There is only limited evidence to suggest that the cell wall plays any role in permitting thermophiles to grow at temperatures that are lethal to most life forms. Differences occur in the rates of peptidoglycan and teichoic acid biosynthesis in *B. stearothermophilus* and *B. coagulans* when grown over a range of temperatures. Both organisms show increased peptidoglycan content and decreased teichoic acid levels in their cell walls, when grown at 55°C, as compared to growth at 37°C. The increased peptidoglycan thickness may serve as a thermal insulator.

Changes in membrane composition are more common in thermophiles. As with organisms growing at low temperatures, the lipid composition of the membrane must be adjusted to achieve a fluid membrane, but in the case of thermophiles the melting point of the membrane lipids must be raised, not lowered. The elevation of the melting point of the bacterial phospholipid bilayer can be achieved by: (a) incorporation of fatty acids with longer acyl chains, i.e. C_{18} instead of C_{16} fatty acids; (b) increasing the degree of saturation in the constituent fatty acid chains; (c) incorporation of branched-chain fatty acids into membrane lipids; (d) increasing the concentration of cyclic fatty acids.

Detailed study of the lipids of thermophilic and mesophilic strains of *Bacillus* and *Clostridium* show that the membranes from thermophiles possess higher melting points and contain greater proportions of saturated and branched-chain fatty acids. The presence of branched-chain fatty acids in the membrane is a characteristic of thermophilic organisms. When grown at 80°C, membranes extracted from *T. aquaticus* contain approximately 60 per cent branched fatty acids (an iso C_{15} and an iso C_{17} fatty acid), but if grown at 50°C the proportion of branched fatty acids falls to about 30 per cent. Bacteria grown at high temperature tend to increase the overall lipid content of the cell. *B. stearothermophilus*, when grown at 60°C contains three to four times the lipid content, as when grown at 40°C.

Microbial membranes contain appreciable amounts of proteins. Growth at elevated temperature causes significant changes in the protein:lipid ratio; in general as the growth temperature is raised the ratio of protein to lipid increases. Many of the membrane proteins are enzymes and studies have been made to determine how thermophilic organisms protect their enzymes from temperatures that would rapidly denature most proteins. Examination of enzymes from a wide variety of thermophiles shows that, in general, the isolated purified enzyme is not significantly more thermostable than equivalent enzymes isolated from mesophiles. It is believed that the enzymes are intimately associated with components of the thermophile membrane and the special nature of this association endows the complex with thermostability.

Although *Thermus aquaticus* grows under conditions that would be impossible for most other life forms, its environment seems tame in comparison with that recently described for a series of methanogenic bacteria isolated from hot volcanic waters. These bacteria are capable of growth at 250°C and 265 atmospheres! Little is known about the structure and physiology of these bacteria and as can be imagined their study provides a severe test for the microbiologist. It will be fascinating over the next few years to learn how such organisms live.

Growth at Extreme pH

Most natural environments are neutral, or nearly so, extreme acidity or alkalinity being toxic to most micro-organisms. Acidophiles have an obligate requirement for an acid environment, with a pH lower than 3, and are relatively uncommon in nature. Alkalophiles, able to grow at a pH greater than 10 are even rarer.

Extreme acid environments are commonly associated with coal-waste heaps and waste-waters pumped from mines. The obligate acidophiles *Thiobacillus thiooxidans* and *T. ferrooxidans* are commonly found in such wastes. *T. thiooxidans* is capable of oxidation of sulphur or sulphide to produce sulphuric acid. The organism grows best at a pH of 2.5 but can survive exposure to pH values close to 0. *T. ferrooxidans* oxidises reduced sulphur compounds, as well as ferrous iron to ferric iron. The oxidation of iron is accompanied by the production of sulphuric acid. The internal pH of this organism (like that of most acidophiles) is maintained at around neutrality, even at low external pH. The envelope of the organism must therefore protect acid-sensitive cytoplasmic components. Examination of the cell envelopes of these organisms has failed to detect any obvious changes related to acid tolerance. Electron micrographs show thiobacilli to possess the typical envelope structure of a Gram-negative organism, with no evidence of any additional surface protective layers. Examination of the peptidoglycan show it to contain no significant changes in chemical composition from that of *E. coli*. However, analysis of the LPS, show that in addition to containing typical LPS carbohydrates (3-deoxy-D-manno-octulosonate, glucose and galactose), the molecule is associated with high levels of calcium, magnesium and in particular iron. These ions may play a role in preserving the stability of the outer membrane in acid environments. The lipid A moiety of *Thiobacillus* appears to possess much lower levels of phosphorus than other Gram-negative bacteria. The absence of phosphorus may render the lipid A more hydrophobic and therefore more stable in an aqueous acid environment. Flagella isolated from *T. thiooxidans* are resistant to both high acidity and high temperature, suggesting that the flagellin produced by this organism is both heat- and acid-stable. The membranes of this bacillus contain a preponderance of amine-containing phospholipids. These may contribute to the maintenance of membrane stability in acid conditions by providing buffering capacity through protonation of their amino groups.

Sulfolobus acidocaldarius, initially isolated from acid hot springs, is a facultative autotrophic sulphur oxidising bacterium. Its optimum growth conditions are extremely severe, 70–80°C at pH 2, although the organism will survive at a pH of 0.9. It is an archaebacterial-type organism and does not synthesise a peptidoglycan layer but a crystalline proteinaceous cell wall. The organism possesses irregular-shaped fimbriae, apparently for adhesion to surfaces such as elemental sulphur. These fimbriae consist of a regular array of protein subunits associated at one end with the cytoplasmic membrane. The fimbriae are extremely acid and heat stable. Treatment of the bacterium with hot detergent releases a pure preparation of cell walls, which can be reversibly dissociated into a single protein species (M_r 140,000–170,000) of unusual amino acid composition. The membranes of *Sulfolobus* are highly unusual, containing 10 per cent neutral lipids, 22 per cent phospholipids and a high concentration of glycolipids (68 per cent). The complex lipids of this organism all contain very long chain, C_{40}, fully saturated isoprenol glycerol diethers rather than typical glycerol fatty acid ester-linked diglycerides. The thermal and acid resistance properties of *Sulfolobus* depend upon an intimate relationship between the protein of the cell wall and the lipids of the membrane. The long C_{40} isoprenoid alkyl chain of the glycerol diether lipids maintains optimum membrane fluidity at high temperatures. The ether bonds of the lipids, which are highly resistant to acid hydrolysis, protect the integral structure of the membrane under acid conditions.

Although not as acid or heat tolerant as *Sulfolobus* the heterotrophic spore-forming rod, *Bacillus acidocaldarius*, grows readily in hot springs with a pH of 2–6 and a temperature below 70°C. Unlike *Sulfolobus* this organism possesses a peptidoglycan-containing cell wall. Membranes of the organism contain the unusual ω-cyclohexyl C_{17} and C_{19} fatty acids, 11-cyclohexyl undecanoic and 13-cyclohexyl tridecanoic acids. These ω-hexyl acids may constitute up to 90 per cent of the total fatty acid composition and are ester and amide linked to complex glyceride-type lipids. The membrane also contains C_{50} and C_{55} polyprenols and a series of terpene-type molecules. In addition *B. acidocaldarius* also contains a sulphonolipid similar to 6-sulphoquinovosyl diacyl glycerol. The unusual and diverse nature of the lipid composition of *B. acidocaldarius* is thought to be important in the survival of the organism in its environment. The ω-cyclo-hexyl fatty acids and the polyprenols may be responsible for ensuring optimum membrane fluidity. Sulphonolipids are highly acid polar lipids and may function in the exclusion of protons.

Thermoplasma acidophilum, found in hot springs of pH 1 to 4 and at temperatures up to 80°C has no demonstrable cell wall. The cell membrane of this organism is directly exposed to a hot acid environment. The membrane contains 10 per cent carbohydrate, 25 per cent lipid and 60 per cent protein. The bulk of the membrane carbohydrate is found in an unusual LPS, (mannose)$_{24}$-glucose glycerol ether, which is thermostable.

The lipid fraction contains mainly C_{40} isoprenol glycerol diethers, structurally related to those of *Sulfolobus*. The phospholipids are not typical phosphatidyl forms, but phosphoglycolipids. The hydrogen ion concentration has a stabilising effect upon *Thermoplasma* membranes; raising the pH results in membrane disorganisation. The *Thermoplasma* membrane is extremely hydrophobic, due to a deficiency of polar groups on membrane proteins and the presence of ether lipids. A low pH provides the ionic conditions necessary to stabilise the membrane. The long hydrocarbon chains of the C_{40} alkyl glycerol diether based complex lipids maintains membrane fluidity at high temperatures, while the diethyl linkage imparts stability to the lipids at low pH.

The occurrence of many unusual lipids in thermo-acidophiles may be of important evolutionary significance. Hot acid environments were probably very common at the time life first developed on earth. The acidic thermophiles may be closely related to the evolutionary precursor of many of today's life forms.

As with acidophiles, alkalophiles, possess an intracellular pH near neutrality. They grow well in alkaline media, with the optimum pH between 11 and 12, although some can grow at pH 12. The most common groups of alkalophiles isolated are Gram-positive spore forming, aerobic rods of the genus *Bacillus*, e.g. *B. alcalophilus* and *B. circulans*. Alkalophilic strains of *Micrococcus* and *Corynebacterium*, *Pseudomonas* and Achromobacteria have also been described. Analysis of the cell walls of an alkalophilic *Bacillus*, when grown at pH 8 and 10 shows that pH has a marked effect on surface components. At the more alkaline pH there are major changes in protein composition, in particular in the composition of the high-molecular-weight proteins, suggesting that protein components may provide an alkali-resistant barrier. Analysis of the cell wall of alkalophilic *Bacillus* grown at alkaline pH (10) shows greater amounts of aspartic acid, glutamic acid and uronic acids, than when grown at neutrality. Alkalophilic organisms may possess larger amounts of those components that endow the surface with a greater net negative charge to protect the cell interior from the lethal high external pH.

Growth in High Solute Concentration (Halophiles)

Many inland lakes in arid areas possess high dissolved salt concentrations, e.g. the Dead Sea contains 32 per cent dissolved salts. The high osmotic pressure of such environments limits the availability of water to any organism present and the number of species able to survive such conditions is low. This is also the basis of the preservation of foods, by salting or sugar syrups.

The extreme halophiles *Halobacterium* and *Halococcus* can grow in saturated sodium chloride solutions (5.2 M), but will not grow below 2 M salt. The envelope of these organisms bears little relationship to other prokaryotes, and in this context the terms Gram-positive or negative are meaningless. The halobacteria, like *Sulfolobus,* lack a peptidoglycan layer, the cytoplasmic membrane being covered by a proteinaceous layer. The halococci, possess thick extremely strong wall-structures, which also do not contain muramic acid, but do contain appreciable amounts of carbohydrates notably glucose and galactose. The extreme halophiles do not contain LPS or lipoprotein. The envelopes of *Halobacterium salinarium* and *Halobacterium halobium* are unique amongst prokaryotes in that they contain glycoprotein. In *H. salinarium,* the glycoprotein is the major surface structure. If it is absent, or its glycosylation blocked by treatment with the antibiotic bacitracin, the cells lose their rod shape and become spherical, suggesting that the glycoprotein is a determinant of cell shape. The envelope lipids of extreme halophiles are unique. In *Halobacterium cutirubium* almost all of the cell lipids are located in the envelope; the main components being diphytanyl ether analogues of phosphatidyl glycerophosphate and a glycolipid sulphate (63 and 23 per cent of the total lipid respectively). The halophile envelope contains only trace amounts of fatty acids, presumably due to the known inhibition of fatty acid synthetase by high salt concentrations. However, the envelope does contain non-ionic lipids in the form of isoprenoid compounds.

Extreme halophiles are therefore characterised by possession of diether lipids and the absence of fatty acids or peptidoglycan. The lipids are usually acidic, containing few basic groups. The acid groups of the envelope protein are located at the cell surface, and presumably interact with the acidic groups of the phospholipids, creating a highly charged surface.

Further Reading

Alexander, M. *Introduction to Soil Microbiology* (John Wiley and Sons, Chichester, England, 1977)

Baross, J.A. and Deming, J.W. 'Growth of 'Black Smoker' Bacteria at Temperatures of at Least 250"', *Nature (London)* (1983), *303*, 423–6

Beachey, E.H. *Bacterial Adherence* (Chapman and Hall, London, 1981)

Bitton, G. and Marshall, K.C. *Adsorption of Microorganisms to Surfaces* (Wiley Interscience, New York, 1980)

Chilton, M.D. 'A Vector for Introducing New Genes into Plants; Agrobacterium Ti plasmids' *Scientific American* (1983), *248*, No. 6, 36-44

Dudman, W.F. 'Role of Surface Polysaccharides in Natural Environments', In I.W. Sutherland (ed.) *Surface Carbohydrates of the Prokaryotic Cell* (Academic Press, London 1977), p. 357–414

Ellwood, D.C., Melling, J. and Rutter, P. *Adherence of Microorganisms to Surfaces* (Academic Press, London, 1979)

Fletcher, M. and Floodgate, G. 'The Adhesion of Bacteria to Solid Surfaces' In R. Fuller and D.W. Lovelock (eds.) (Academic Press, London and New York, 1976)

Gray, T.R.G. and Postgate, J.R. *The Survival of Vegetative Microbes* (Cambridge University Press, 1976)

Heinrich, M.R. (ed.) *Extreme Environments; Mechanisms of Microbial Adaptation* (Academic Press, London and New York, 1976)

Kushner, D.J. (ed.) *Microbial Life in Extreme Environments* (Academic Press, London and New York, 1978)

Marshall, K.C. 'Growth at Interfaces', In M. Silo (ed.) Dahlem workshop on *Strategy of Life in Extreme Environments* (Chemie Verlag, Berlin, 1979)

Poindexter, J.S. 'The Caulobacters; Ubiquitous Unusual Bacteria', *Microbiological Reviews* (1981), *45*, 123–79

Rheinheimer, G. *Aquatic Microbiology* (John Wiley and Sons, Chichester, England, 1980)

Sleytr, U.B. and Messner, P. 'Crystalline Layers on Bacteria', *Annual Review of Microbiology* (1983), *37*, 311–19

Sprent, J.I. *'Biology of Nitrogen-fixing Organisms* (McGraw-Hill, London, 1979)

Woese, C.R. 'Archebacteria' Scientific American (1981), *244*, No. 6 94–106

INDEX